NEVER GIVE UP

Mission Raptor
Mission Jaguar
Mission Gobi
Mission Alps
Mission Wrath
Mission Tsunami
Mission Volta

Bear Grylls Colouring Books: In the Jungle
Bear Grylls Colouring Books: Reptiles and Amphibians
Bear Grylls Sticker Activity: Dangerous Animals
Bear Grylls Sticker Activity: Animal Detective
Bear Grylls Sticker Activity: Wild Survival
Bear Grylls Sticker Activity: Polar Worlds
Bear Grylls Sticker Activity: Beastly Bugs
Bear Grylls Sticker Activity: Amazing Birds
Bear Grylls Sticker Activity: Extreme Animals
Bear Grylls Sticker Activity: Endangered Animals
Bear Grylls Sticker Activity: Predators
Bear Grylls Sticker Activity: Reptiles and Amphibians
Bear Grylls Sticker Activity: Deserts
Bear Grylls Sticker Activity: Rainforests
Bear Grylls Sticker Activity: Woodlands
Bear Grylls Sticker Activity: Mountains
Bear Grylls Sticker Activity: Coasts
Bear Grylls Sticker Activity: Under the Sea

Bear Grylls World Adventure Survival Camp
Bear Grylls Extreme Planet
Bear Grylls Epic Adventure Series: Epic Expeditions
Bear Grylls Epic Adventure Series: Epic Climbs
Bear Grylls Epic Adventure Series: Epic Voyages
Bear Grylls Epic Adventure Series: Epic Flights

A Bear Grylls Adventure 1: The Blizzard Challenge
A Bear Grylls Adventure 2: The Desert Challenge
A Bear Grylls Adventure 3: The Jungle Challenge

NEVER GIVE UP

BEAR GRYLLS

BANTAM PRESS

TRANSWORLD PUBLISHERS

Penguin Random House, One Embassy Gardens,
8 Viaduct Gardens, London SW11 7BW
www.penguin.co.uk

Transworld is part of the Penguin Random House group of companies
whose addresses can be found at global.penguinrandomhouse.com

Penguin
Random House
UK

First published in Great Britain in 2021 by Bantam Press
an imprint of Transworld Publishers

A CIP catalogue record for this book
is available from the British Library.

ISBNs 9781787634190 (cased)
9781787634206 (tpb)

Set in 11.5/14.75 pt Minion Pro
Typeset by Jouve (UK), Milton Keynes
Printed and bound in Great Britain by Clays Ltd, Elcograf S.p.A.

The authorized representative in the EEA is Penguin Random House Ireland,
Morrison Chambers, 32 Nassau Street, Dublin D02 YH68.

Penguin Random House is committed to a sustainable
future for our business, our readers and our planet. This book
is made from Forest Stewardship Council® certified paper.

To my amazing Shara – the kind, smart,
loving bedrock of our family.

&

To Delbert Shoopman III and Rupert Tate for your friendship –
Warriors beside me on this journey.

CONTENTS

CONTENTS

CONTENTS

PROLOGUE

I CHECK MY watch again. 0933. People say time can stand still. It's not true. But it sure as hell feels like it sometimes.

I have been awake since 4 a.m. And two hours' sleep is never enough. It's been a mission just to get here for sure. I am tired. But I'm also adrenalized.

I look around me. I see several of the bushes rustling. I know who is hidden in there, and smile. I know they are as nervous as me. Our small team of eight. Small but mighty. Carefully positioned throughout the trees. They will emerge with no fuss and join me at the pre-planned moment.

But for now I am alone. And I feel it.

I have to remind myself that somewhere within 200 metres we are also joined by a ton of highly trained snipers and close protection officers – none of whom I can spot. I guess that's their job. It's not a comforting thought, though.

Breathe, Bear. All is good.

The huge Alaskan mountains tower above us. They don't seem to care much about all of this. One of the many things I love about mountains – they never judge, never praise, and never seem to worry. They just are. And they are amazing. A good lesson to remind myself of.

Come on, Bear. You got this. Breathe it in.

The cool summer air feels good. And the mosquitoes have now gone. I hate mosquitoes. God knows they have had their fair share of my flesh over the years.

I try and rehearse the first few moments in my mind. The interaction. The questions. The route. Another huge military helicopter sweeps along this vast valley, hugging the edge of the glacial moraine. A last security sweep of the area, I figure.

I take another deep breath. The words of my mother ring in my ear: 'You only get one chance to make a good first impression.'

Don't screw it up.

I reach down and pick up a few stones in my hand and shuffle them round and round in my palms. *Calm is contagious.* One of the mottos of the Navy SEALs. I know it well. Not always easy to practise though.

Come on. Where the hell are you?

I look left and right.

Nothing.

'Late is where you are . . . and that would be press-ups,' I mumble to myself out loud.

No one can hear me. Except Jimmy, our sound man, via my remote mic. He will be smiling at that from his position further along the river bank. Hidden. He is one of the few among our crew who laugh at my bad jokes. I love him for it.

Yep. Press-ups would be being issued right now.

During my army days, press-ups were always the standard punishment for the troop if any of us were late on parade. I wonder if that rule applies if you are the Commander in Chief of the largest, most powerful military ever known to man. Probably not.

I check my watch once more. Still 1053.

My radio crackles with static. It makes my heart jump.

And then, suddenly, everything goes strangely still.

I can feel it.

Here we go.

It's game time.

Then, with no fanfare or fuss, and flanked by ten Secret Service agents clad in black, weapons ready, the President of the United States emerges from the trees.

And I am here at his request.

How on earth did that happen?

1

MUD, SWEAT AND FEARS

THE TRUTH ABOUT reaching the top of any mountain is that the only way forward is down.

That is a big part of why I have taken so long to write a follow-up to *Mud, Sweat and Tears*, the autobiography I published in 2011. Its success surprised my entire family. And me too. It sat at the top of the *Sunday Times* Bestseller List for over ten weeks. Before then, I had never been number one at anything.

When in 2012 the book was voted the most influential in all of China I felt that we were done. Good job, team. We'll order a few pina coladas (as is customary in our crew), a mountain of pizza (always), then tomorrow we get back to work. On to new territories. New mountains.

Remember: sit on the summit too long and you die. A fundamental Everest lesson. And from the top, as I said, the only way is down.

Some people might feed off success. I don't. I actually struggle with it. It feels indulgent. It makes people slow. I see it often. And I am all too aware of the reality of the many factors that allow success to happen.

My own overriding feeling is always: there by the grace of God go I. And I understand the huge role that luck has played in my career. Of that I have no doubt. Then again, luck's a strange thing. As one of my heroes Ranulph Fiennes once said: 'Luck is only half of the story.

You've above all got to have the resilience to stick around long enough for the luck to come.'

I've always liked that.

Anyway, either way – luck or determination – I was always told as a kid to quit when ahead, and to leave parties five minutes too early, rather than too late. That attitude was ingrained in me from a young age. Really my parents were telling me: Don't be greedy, always be grateful.

For that reason I was always hesitant to write a sequel to *Mud, Sweat and Tears*. I wanted to let it be. Call it fear if you like. Or fears, more accurately, which I've most definitely gathered over the intervening years.

But then fear is a terrible reason not to do something. And just because a mountain is big, it doesn't mean you shouldn't climb it. Then there is the fact that some stories simply ache to be told. And there is no doubt the last ten years of my life have been the most extraordinary any adventurer could ever imagine.

So, screw it. Here we go . . .

2

MAN VS. WILD

ONE OF THE things I'm most often asked about is *Man vs. Wild*, and the many adventures we filmed on that show. I didn't tell those stories in *Mud, Sweat and Tears* because I felt there was a lot of other stuff to say: the unknown stories that shaped who I was when I was growing up, and as a young man trying to set out on the adventure of life. Whether it was surviving school, or joining the military, or breaking my back in that free-fall accident and then climbing Everest. Our first transatlantic Arctic rigid inflatable boat (RIB) expedition, or paramotoring above the Himalayas, and the many other expeditions that have shaped my life, and almost ended it on way too many occasions. Those felt more like the real cornerstones of my life at that time.

But there's also no doubt that *Man vs. Wild* was the door opener to so much else that has happened since. To have been given a chance to do primetime US television on a global network for a channel with the reach of Discovery was a huge privilege. You often hear British actors talking about wanting to break America, but that road is littered with failed attempts. I was one of the fortunate ones.

In truth, at the time I never really appreciated that good fortune and the incredible opportunity I had been given. I was always more excited about doing Channel 4 in the UK than some distant American station – mainly because my mum would see it. I was so naive.

For that golden US opportunity I have a British producer, Rob MacIver, and Diverse Productions to thank. I'll always be grateful for Rob's faith in a scruffy former soldier who kept turning down his offer of doing an American TV show. I guess it was a lack of confidence in myself as much as it was fear of the unknown, but I just figured TV wasn't for me. So I said no to Rob three times.

Rob believed in me when I had no idea what I was doing. To this day he still says that I was rough and clueless, but that he knew that if we could just film some adventures together, and if they were as wild and fun as we reckoned they could be, then American audiences would love it.

I had no idea if he was right, but eventually I went along with it anyway. Rob's US TV idea, *Ultimate Survival*, as the first incarnation of *Man vs. Wild* was called, paid better than the Channel 4 series I had just done called *Escape to the Legion*. So I didn't have much to lose, as my wife Shara pointed out pretty wisely.

That first *Man vs. Wild* series changed everything.

Looking back now, two things are crystal clear: first, that the power of being promoted so heavily year after year on primetime American TV was the key to so much that followed in my life. And second, in making that show we had some truly mind-blowing moments that will stay with me for ever.

And that's where I want to take you.

I could begin with the self-induced enema on that raft I had made off some Pacific desert island; or that log that snapped under me while crossing a 100ft ravine in Alaska; or the pit viper that bit me in the Borneo jungle; or the catfish noodling in the alligator-infested swamps of Alabama. Or maybe one of the close encounters with saltwater crocs in the Australian Northern Territory, or being buried in an avalanche or caught in a rockfall. Perhaps the improvised wetsuit I made from a rotting seal carcass, or the urine-drinking, faeces-munching, porcupine-hunting, tarantula-chewing, river rapids-running, mountain-climbing, free-falling, water-landing moments . . . of which there were so many.

And so many questions, too. Such as the worst thing I ever ate. Or my favourite *Man vs. Wild* episode. Or the toughest place I have ever had to survive. Or the worst.

Questions that feel like they need answering.

So, for the sake of some small indulgences, let's do it: *Man vs. Wild*, from my perspective. Just for the record.

3

WILDERNESS HYDRATION

I'VE BEEN ASKED more times than is imaginable, do I really drink my own urine? I don't know what it is with folk. Disbelief, mostly, and an insatiable appetite to know what urine tastes like.

Well, the answer is yes I have, and yes it is terrible. But no, definitely no, I do *not* drink urine for fun. Or for health. Although there are some people out there who seem to swear by it. I'm not one of them, but I have drunk it a bunch of times in the name of survival. And the answer to the question 'What's it like?' is: 'It stinks.'

Warm, salty urine is not designed to taste good. Especially if it's been stored in the skin of a rattlesnake while crossing a desert. Mixed with blood and snake innards, that urine took on a taste of its own that I'm not keen to repeat in a hurry. Then again, survival rarely tastes good and almost always hurts, stinks and leaves you a little beaten up. That's the reality of the wild for you, and most certainly the reality of survival. It can make you suffer. But there's part of me that loves that.

Life can be so sanitized nowadays. We shy away from struggle and we reject the broken, the fallen, the unconventional and the 'not fit for purpose'. For the survivor, and for humankind over the millennia, those very things mean opportunity. Good survival means thinking left field, digging deep, doing the unimaginable, and yes it might hurt and will probably stink. But in terms of staying alive, the rewards always go to the person who can dig the deepest and find that something inside that allows them to do the unimaginable. That ethos became *Man vs. Wild*.

The few survival shows that had been on television before had always generally been about taking minimal risks, getting set with a shelter, lighting a fire and waiting for rescue. To me, those were the most boring parts of survival. I liked the constant movement, the relentless pursuit of escape. What would you do if you knew no one was coming to find you and you had to move, you had to self-rescue? Oh, and what if wolves were also on your trail, for good measure? You've got no tools, just your wits and your survivor spirit to drive you on . . .

Boom! Now we have an adventure on our hands. And that was the show: *What if* . . .

So, anyway, yes, urine tastes bad, but it can save your life if you are well hydrated. *Don't waste clear pee* is the principle, because in a survival situation staying hydrated is right up there in terms of your priorities. Without precious fluids you go downhill fast. So conserve, be smart, do the difficult, drink the clear urine, survive. (But remember, if your pee's brown and stinking that's nature's way of telling you to stay clear. Pure waste product.)

I always remember when our eldest son Jesse was about eight, and I was doing a workout outside, smashing through some burpees and press-ups, clock ticking, high intensity, water bottle beside me, working at my max. It's always the way I like my workouts – fast body weight circuits that build speed, strength and flexibility. Functional fitness that helps me do my job.

Anyway, Jesse walked past and I asked him between gulps of air to do me a favour: to quickly refill my water bottle. He picked up my empty one and scurried inside, returning with the bottle . . . and a smile.

I finished my set, hands on hips, sucking in the air for my ten seconds' rest before the next exercise.

'Thanks, champ,' I mumbled as I took the water bottle.

I unscrewed the cap and drank deeply . . . so good.

So . . . salty! I spat it out violently.

Instead of water, I had taken a mouthful of Jesse's warm pee.

'What? Why?'

'But you love drinking pee, Papa!'

I guess that urine gag is going to follow me around for a while.

It's not a great drink, I have to admit. But what about the worst thing I have ever eaten?

It would be tempting to say those camel intestinal fluids, or the frozen Siberian yak eyeball full of blood, fluid and gristle. Then again that New Zealand giant weta insect was off the scale. Or maybe it would be live scorpions, which are always terrible – full of some weird yellow goo; or the elephant dung, or the berries scraped out of bear faeces that had a particular tang for sure. Skunk anus and rat brain were low points. But they all pale in comparison to that raw, swollen goat's testicle I once ate in the Sahara . . .

4

BEWARE OF TESTICLES

I HAVE LEARNT from bitter experience, after many years of trying local delicacies . . . to be particularly wary of trying local delicacies. Whether it is a local hooch made from the fermented saliva of the Emberá tribe in a Panama jungle, or pickled chicken feet in China. Experience has made me cautious. Very cautious.

I wasn't always so astute. That testicle from that particular goat in the Sahara was a low point.

We were spending the night with the local Berber desert dwellers in their makeshift camp amid the sands in a very remote corner of Western Sahara. They were adamant that they wanted to dispatch a goat and that we would all feast together under the stars. Their wonderful sense of hospitality was hard to refuse.

Oh, and I would be offered one of the goat's testicles as an honour, as their guest. That meant I would get the privilege of eating it warm and raw. The prized, precious private parts, so to speak.

The crew's eyes lit up. God, I was green.

The Berber headman ushered me behind the tents to where said goat was tied up. The goat wasn't so big. *A good sign*, I thought to myself. I figured that while the Berber was busy sharpening his blade I would just check out the size of this goat's balls. Mental preparation and all.

I wandered slowly around the back and bent down for a look-see.

13

Mother love. They were huge. In fact out of all proportion, it seemed. This fella had melons slung under him. It was truly Sod's Law that this was the best-hung goat in the entire Sahara desert.

The knife was sharp and the Berber slit the poor goat's throat in a second. It's a part of life for these folk that I wasn't going to question. Especially this far out in the desert.

His companion collected the blood as it drained from the cut and soon the goat was hanging upside down ready to be prepared for the feast. The testicles hung down like saggy, swollen water balloons.

A few deft swipes of the blade and the sack was open, with the first testicle severed and lolling like a giant white jelly in the Berber's hand. The size of it was only rivalled by the size of the grin on his face. Head nodding at me, hand outstretched . . . and cameras rolling.

For me, there has always been something about the camera's blinking red light that helps me do the difficult. It means we're on, we're green to go, and it's time to deliver.

I always loved the Scouting motto: *Do your best*. DYB DYB DYB. Summed up always for me by my mother's words: 'When a job is once begun, leave it not 'til it is done; be it big or be it small, do it well or not at all.'

And on this occasion the job was big.

I reached out, and the ball slid into my hand. It was heavy. Wow.

A deep breath. *Do the difficult. Do your best.*

'Uno, dos, tres, in we go.'

I squeezed the testicle into my wide open mouth and chewed with a grimace. The ball seemed to hover then suddenly explode into a mouthful of what I could only presume was goat sperm, and I choked to swallow it down. But the gag reflex and pungent taste was so strong that I instantly retched and felt the vomit rise within me. I tried to counter the gag and get the sperm and sack down my throat.

Come on, I urged myself.

The vomit, though, burst into my mouth and swirled in with the jelly.

Hold it in, Bear.

Even the Berber was grimacing. Our eyes met for an instant. And sometimes a look can transcend a lack of language. His eyes screamed one thing: *That was one mother lode of a big testicle, eh?*

I finally got it down, vomit and all. Job done. I even managed a very British apology to him for the fuss. He smiled, then laughed out loud.

Camera off. It had been a long, hot day for us all. Moving on, team. Moving on.

So yes, that testicle was bad. But if you were to ask me the toughest place I've had to survive . . . well, that was a world away from the Sahara, in one of the coldest, grimmest locations on Earth.

Siberia. In winter.

5

TOUGHEST

IF YOU HAD to pick one place on our planet that is brutally unforgiving in its harsh, cold, desolate conditions, then Siberia in December would definitely be in your top three. Probably top two.

It is a long, long way from anywhere, difficult to get to and, as a survivor, an equally hard place to get out of. It took me to the edge.

It's always special, and sometimes quite eye-opening, travelling to the really remote parts of the world – even more so when in a country that has been under the shadow of communism for so long. Where much of Russia changed very fast after the fall of communism, Siberia moved a little slower. In a world where the average temperature in winter is a balmy minus 25°C, everything takes longer.

Four flights in planes of ever-decreasing size, and ever-decreasing reliability, and we find ourselves in tiny log cabins, each the size of one and a half table-tennis tops, with a single bed and a small iron stove for warmth, in the middle of the taiga forest in deepest Siberia. For the six years we shot *Man vs. Wild*, this sort of situation was not uncommon. Just another difficult place to get to, with locals who were essentially surprised at what we were proposing to do.

Our team generally comprised fewer than ten people. A mix of camera operators, sound recordists, a mountain guide or two (to help keep crew safe), a medic, a logistics coordinator and a producer, plus a local fixer who would help sort stuff out, if and when things went wrong. Again, this was not uncommon.

An initial briefing for the whole team from local search and rescue teams or the local rangers would often be a sobering affair, reminding us that the surrounding wilderness can be truly unforgiving (especially true of Siberia at this time of year) – warnings I became all too desensitized to hearing. In short, *Man vs. Wild* locations and conditions were rarely easy.

We had some easier moments for sure, and often at the end of twelve days charging through these jungles, deserts, swamps and mountains we would end up at a half-decent place for a night. But for the most part, it was hard. I just didn't realize it so much at the time. Nowadays, the crew and I look back with some head shaking and smiles at the pace we moved at, the places we went and the stuff we got up to. I guess we were learning our trade, testing our mettle and doing our time.

Anyway, shoots would always begin before dawn, and when you are in conditions where the temperatures are either super cold or super hot, everything very quickly becomes hard work.

We became masters of managing ourselves in tough environments, and that was nothing to do with what we were actually filming. That was just about maintaining yourself, conserving yourself, pacing yourself to make it through the day. It was often simply about surviving through the day's adventures, to reach the next short night, in order to be ready for the next ball-buster day ahead.

And repeat . . .

By the end of six seasons we had *Man vs. Wild* shoots down to five or six days, but in those early days of back-to-back filming, shooting several shows at a time, it was a brutal schedule where pacing yourself through the crazy hours and intensity was key. And much of that endurance was about a state of mind.

When I look back on the sort of things we would be doing all day, every day, it seems crazy. But iron sharpens iron, and being sur-rounded by the crew I filmed with, a group of legends, ensured we always got the job done. We kept each other going and kept each other laughing through so much, taking the mick out of each other dawn to dusk. Through the fears, the cold, the heat, the long treks, the heavy packs and all the blood-soaked, stomach-churning moments that made the show, the crew were best friends. And still are.

To this day, many of them are still the ones beside me on our Net-flix, Amazon Prime, ITV and National Geographic shows – still there, still laughing, still teasing, still yomping the hills and carrying the heavy packs. So when I call them legends, it's not a throwaway line. They truly are. Unsung by an audience, but heroes to me. The real ones.*

---

* You know who you are: Mungo, both Dans, James, Simon, Pete, Paul, Dave, Duncan, Danny, Jimmy, Scott, Dilwyn, Josh, Woody, Matt, Jimmy, Ben, Rob, Jeff, Ross, Meg, Stani and Liz . . . I love you guys and am so grateful for all you do.

6

DOING OUR TIME

ANYWAY, BACK TO Siberia. I remember one freezing Siberian night, cameraman Simon Reay and I were curled up in our sleeping bags on thin roll mats, laid on the deep snow under a tree. Neither of us were sleeping, and every hour or so he would prod me to show how the temperature dial on his camera bag had dropped even further.

'I know it's cold, Simon. I've got snot frozen to my nose and it hurts to breathe.'

His face would be squashed through a small gap in his bag, exposing just the tip of his nose and his grinning lips.

'Minus forty-two. That's epic.'

It was as if the crazier the situation, the more we knew as a crew we were doing something special. I can still see the look of pride on Simon's face as, shooting the next day, water froze in a matter of seconds on the lens of his underwater camera housing.

'Shit. You see that, guys?' he shouted. 'Crazy cold. Literally froze as I wiped it.'

Dave Pearce, our safety expert, ex-Commando and all-round good guy, is holding Simon on a short rope, bracing him against the flow of icy water upstream of me, as he edges out into the river.

Meanwhile, I am now naked in minus 35°C, trying to smash through the thin ice at the edges in bare feet, in an effort to reach the main flow – the black, fast-flowing middle section – and eventually across to the far bank. I am shouting at Simon to keep with me and keep moving. Camera or no camera, we aren't stopping now.

Move fast. Don't hang around. It was the DNA of our time. Especially in the extremes of temperatures. It was where the whole notion of 'no second takes' for anything came from. Pure necessity. Get it or don't. We're not doing that again. Moving on.

Meanwhile the rest of the crew are already on the far bank, stomping their feet and rubbing their hands in an effort to keep warm, buried under all their layers – just as I had been until three minutes earlier when I knew it was time to strip off and do the job.

Both Dave and Simon are dressed in full dry suits. Both are still freezing cold. We all are. But, naked, my survival time is getting really low, especially if I find myself in trouble in this river.

Often, I have seen people freeze in shock at their first ice water immersion, and be unable even to say their name, let alone swim naked, break ice, climb a deep snow bank and then get a fire going with only a fire steel and flint. But this sort of thing was so typical of what *Man vs. Wild* embodied. Adventure survival.

Man vs. Wild became a study of getting into just enough 'planned trouble' without getting caught out by the unplanned stuff. It didn't always work that way.

If all went perfectly to plan, then the swim-jump-crossing-submersion in front of us was possible, but what became clearer and clearer as time went on was that if things didn't go to plan, we would be in serious trouble. In other words, the margin for screw-ups was small. That was where the danger came in.

Such as the time that Alaskan fishing trawler knocked me off my feet as I balanced on a small iceberg, hoping to grab hold of a thin rope ladder lowered from the ship's side. I only just caught the rope and avoided being crushed and sliced in two between trawler and berg.

And there was the high gorge crossing when that log gave way beneath me and I just managed to catch another branch with one hand; or the rockfall that sent a boulder the size of a car flying between Dan Etheridge, our cameraman, and me at 100mph, only missing us by feet.

The lesson was: don't play the odds for too long, and what you can plan, plan well, do once and don't screw up. Or else. It's in the 'or else' part that Mother Nature can be very unforgiving. And when things go wrong, they tend to go wrong fast. As they say: nature is like your momma . . . respect her and she will treat you right; disrespect her and she'll teach you a lesson you won't ever forget.

Anyway, the Siberian river crossing went well. We didn't die, we all delivered on our jobs, Simon caught the action, Dave kept Simon safe, the microphone worked despite the water, and so on. As the Royal Navy say: the team works. But we never had much time to take stock and say well done to each other. The pace was always so fast. Twelve days in a country and two episodes to nail. Keep going. So much to do.

Once out of that river it was on to digging a snow cave, making snares, climbing cliffs, sledging down huge 1,000ft snow faces on an animal carcass and drinking fresh blood from a yak's neck. Then chasing down the trans-Siberian express train and skydiving from an old Russian Mi-17 rust bucket into a white-out. (We would never have got permission for a jump in weather conditions like that in the UK. No way. But in Siberia . . .)

The list of crazy stuff was unending on *Man vs. Wild*. Every day different. Every day a close call. It was a wild ride but it was just part of the job. We rarely questioned it. I considered it as simply doing our time.

7

SLAV THE IMPALER

MY LAST MEMORY from Siberia is sleeping in a cave with Dave, on another ballistically cold night. The rest of the crew had an hour's drive in an old Russian truck back along the narrow mountain forest lanes to those log cabins and the generator, where they could recharge camera batteries, get some hot food and some rest before a 4 a.m. start to come and rejoin us the next day.

That moment when the crew departed with the cameras, and filming was done for the day, was always my favourite part. I have never been someone who likes being the centre of attention. Still today, cameras to me are a necessary evil to be endured to enable me to do these adventures with great friends in wild places. But that's the job. And back then it was my only job – and only source of income. I had to keep doing it. Embrace the difficult. Know that the cameras will stop rolling at the end of the day.

Because being on camera was something I struggled with. At the start of every day I would always have to rediscover the motivation – not to do the adventure (that was the part that felt natural to me) but to put myself in front of that lens. I really wrestled with it. I just didn't like people watching, critiquing and debating my every move, which in the early days was invariably what would happen.

I would make a trap, then the producer, and whoever else was around, would discuss in hushed voices whether it was good enough or whether it needed to be done again. I hated that waiting. Especially as my brand of survival was always a little rough – good enough is good enough, as I used to say. This wasn't a spoon-whittling lesson, but rather a demonstration of how to get your backside out of trouble, and fast.

For that reason, pretty early on I said to the team I couldn't do it that way. The retakes and endless scrutiny were killing me and the spirit of it all. And TV is so revealing that if you lose enthusiasm, it shows. As soon as I felt that happening, I said that if this whole thing was going to have any legs at all we would have to do things once, keep it natural, allow me to ignore the camera as much as possible, and the crew would have to do all they could to cover it. Get it or don't. Move on.

The producer of the time pushed back – it's impossible, we won't get the coverage, the close-ups, the cutaways. The show will be a total failure. I held my ground. And guess what? The magic returned.

I had slowly found a way, through necessity, to overcome that fear.

Just blot them out, Bear . . . Focus on the adventure. And do that part well.

Today, I am much more comfortable with the whole filming thing, but only because I have learnt to ignore the cameras and just do the adventure stuff without thinking about the lens's scrutiny. It's why we now have about four cameras minimum on every shoot, not one. Everyone knows that in order to get adequate coverage of a scene it takes multiple cameras, always rolling. Even if the lens is covered in mud and rain, it's OK. Keep it rough. It works.

At least now you can understand why I always loved that end-of-day 'cameras off' moment on *Man vs. Wild*. I could sit down on a log for the first time in twelve hours and just stop. No cameras. Relax.

That Siberian evening, the crew left, and Dave and I stoked the fire and prepared for a cold night in the cave, 300ft up at the base of a steep cliff.

We looked down at the track junction below us, where Slav, our local guide, was going to be sleeping. Slav hardly ever spoke, weighed in at about 23 stone, and only seemed to eat smoked pig fat and gristle, which he impaled on his trusty Russian penknife. And instead of tea I only ever saw him drink Siberian vodka. We thought he was brilliant.

'Maybe we should just check he's OK before we get in our bags for the night – what do you reckon?' Dave asked me.

We wandered down to his camp to find him keeled over on his back, midway between the fire and his tent. He was snoring. The empty bottle of vodka told the story.

'The guy will die if we leave him here in these temperatures,' I said to Dave.

We tried to prod him awake, but nothing. We tried to shift him into his tent. The guy wouldn't budge.

It took both of us, and all of our strength, to finally lever him the few metres to his old canvas tent and push him inside.

'You think he'll be OK?' I asked.

'Siberian tough. He'll be fine.'

That night was even colder than the last, when Simon had stayed out with me. Dave and I burnt through a pile of wood the size of a small house during those dark hours. Neither of us slept more than twenty minutes at a time, before one of us would have to get up and stoke the fire with more logs. Without that fire, that cold cave would have devoured us.

By dawn, when we emerged into the cold mountain air and went down the slope to check on Slav and to wait for the crew, we were a little anxious about what state he would be in. And we could see that his fire had long gone out. In fact, I was genuinely concerned if he would even be alive.

As we rounded his tent I wondered why we had ever even worried. There was Slav, grinning at us, dressed exactly as we had left him: big thick grey overcoat, with hands like spades pushed through the sleeves, chewing on a piece of impaled pig fat on the end of his knife.

Dave and I looked at each other and laughed. It would take more than a minus 45°C night to finish off Slav the Pig Impaler, as Dave started to call him. Wow, they make 'em tough in Siberia.

Nowadays, whenever I am having a moment of self-pity or complaining about the conditions or the terrain or the journey time, I often think of Slav. Hard as effing nails, hands like boats, and always smiling – whatever the conditions.

Slav the Impaler. Legend.

8

ONE TEAM, ONE MISSION

DESPITE DRAWING STRENGTH from characters like Slav, one thing that it is important to make clear from those *Man vs. Wild* days is that, at almost every turn, if I ever wanted it or asked for it, I would get as much help as I needed from our team.

Despite how people liked to perceive the show (and despite the disclaimer at the start saying 'Bear Grylls and the crew receive support when they are in potentially life-threatening situations'), I never saw *Man vs. Wild* as one man against the elements. We were a tight band of brothers (and sometimes sisters), and the huge support and many unseen kindnesses and helping hands that team gave me, every day, were what made that show possible.

It's as simple and truthful as saying that without that team, *Man vs. Wild* would never have happened. It is why I never really liked the attention and sometimes over-fantasized praise that *Man vs. Wild* brought. I could never be as strong and capable as the show would often make out: one man battling alone and always winning, against overwhelming odds. That's not how life or the wild is. You might survive once like that but not every week, every year. It takes a team, albeit a small one, but above all a great one, to make that sort of stuff happen safely and consistently.

I have always wanted people to know how much our team helped me on every mission. That without them I would have died many times over. But whenever I would talk about that sort of thing on press days, people never seemed to care much about those less dramatic realities. Their readers or viewers wanted the danger and the battles and the crazy stuff. It was always about that headline, or those two minutes of a story on a chat show couch, and then they were done. That was what the journalists cared about. The hero image fitted their bill. But I knew the reality and I always valued, more than anything, the power of that team. Without those real heroes behind or beside me, every step of the way, working harder, longer, and often doing a more intricate, skilful job than mine, I would never have 'survived'. In more ways than one.

Although, now I think about it, there was that one time in the Sahara desert when it very nearly did become *Man vs. Wild* – one man against it all. And it still makes me smile.

We really should have known better: the height of summer in the mid Sahara is not a good time to film. Every day, up and down sand dunes, climbing, clawing, rolling, hauling, making, building, all in 50°C heat – that was always going to reduce our already slim margins for error.

Sure enough, soon one man went down with heat stroke and vomiting. That meant our first crew vehicle and a driver gone, called upon to evacuate him from the searing heat. Then the medic went down, which meant the second vehicle and another driver gone, across the dunes back towards civilization.

The day got hotter, but we kept going, kept filming. We always did. Come midday we lost another one, plus our second-to-last vehicle. We were now down to just me, Simon the cameraman, Paul Ritz our sound recordist, and a brilliant former Paratrooper called Danny Cane, who always kept the shoots working, in terms of logistics and timings.

This was now a bare-bones crew, and with just one car and driver remaining, we kept to plan, trying to follow the camels we were tracking on foot across the blistering dunes in oven-hot temperatures, where just to breathe hurt your throat.

And then finally, Paul, who was unstoppable normally, went to one knee, threw up and started to wobble.

'No, I'm OK, I can keep going,' he mumbled as he soldiered back to his feet.

But twenty minutes later it was clear he wasn't going any further. We helped him to the final car, poured water down him and put the air con on. He would need to get back to base fast. Heat stroke can be genuinely dangerous, and we all knew this. We had already had reports that the earlier heatstroke cases were more serious than we had thought. One of the team was now in hospital.

We had to make a call about whether to down tools or press on. We just had one more segment to film to complete the show and wrap it up. We had to nail it, and all agreed to stay for one last push. But how could we manage without a sound recordist?

Now you have to understand the charade that Paul had given us over multiple years and multiple adventures about how intrinsically complicated being a sound recordist is – how they are the only really indispensable person on a shoot and that us monkeys wouldn't even know where to begin to do his job. We had all heard the story many times.

But this was a desperate time and it called for desperate measures. What could we do? The only man still standing who could possibly replace Paul was Danny, the ex-Para. But Danny, for all his many skills, could hardly even work a mobile phone let alone possibly know how to operate a sound recordist's mobile mixing desk, with all the dials, needles, fluffies and microphone boom poles.

Paul was now lying slumped in a heap, intermittently throwing up, faint with heat stroke. He knew his time was nearly up.

'Paul, either tell us the truth about this sound engineer business or we fail in our mission,' I implored him.

And it was then, at the point of utter desperation, that we first heard the immortal words about how to be a sound recordist.

'Danny, come over here,' Paul mumbled, like Nelson summoning Hardy for his farewell kiss. 'Now, listen carefully . . .' We all held our breath. 'Basically,' Paul continued, 'it's all quite simple: just keep the boom out of shot and the needles in the middle.'

And with that he passed out.

Ever since then, whenever a soundie starts getting technical with any of us on the team, or complains that he needs time to sort out some acoustic issue, while we are in some jungle gorge trying to do our thing, we all just look at each other and in unison will say: 'Why? All you have to do is keep the bloody boom out of shot and the needles in the middle.'

Genius.

Anyway, Sahara done. Survived. Moving on.

9

APOCALYPSE SURVIVAL

IF YOU WERE to ask me to choose one episode I loved doing more than any other, it would be the urban one we filmed in an old disused marine dockyard on the Baltic coast of Poland. We were on season five of *Man vs. Wild* by this stage and keen to keep mixing things up and taking adventure and dynamic survival to the next level. And lots of folk had suggested an apocalypse-type scenario.

As a team, it was epic to shoot. We had the space and freedom to break some of our own rules and the conventional *Man vs. Wild* formula, and to go all-out apocalypse.

The base we shot in was a former Nazi slave-labour dockyard historically used to repair U-boat submarines – and it had hardly been touched since. A vast expanse of huge old warehouses and hidden tunnels, with fuel dumps, oil reservoirs and broken glass everywhere. The place was full of dust, grease, pigeon crap and a sense of history and hardship that was difficult not to feel.

The episode started screaming in on a fast Special Forces-type RIB craft and leaping for dear life into a cargo net suspended from an old crane over the dockside. And from there it was a blast. Up cables, down elevator shafts, across electricity wires, even blowing off doors to gain access to a hidden chamber by igniting old oxygen and acetylene bottles.

It was a fun chance to make the whole thing feel a bit more like a movie, with some cool elements, such as doing parkour and back-flips off rooftops, swinging in through windows and getting wedged in narrow ventilator shafts. It meant we had to 'produce' the show a bit more and plan it carefully, but what it lost in the spontaneity of charging down scree slopes and crossing rivers we gained in a more scenario-type survival.

Anyway, we all loved it, it broke the mould, and we did something totally new with a well-trodden format. I was so grubby at the end of filming, caked in grease and oil and grime, that it took a whole week to look normal again. And I have generally set a pretty high bench-mark for getting dirty.

In fact, it became a bit of a joke at home that when I returned from filming *Man vs. Wild*, Shara would ban me from entering the house until the bags were emptied and I had stripped off and hosed down. The sort of things that would tumble out of those bags, backpacks, my pockets and my ears really should have been catalogued. From rat tails and rattlesnake fangs to burnt-out flares and blood-soaked socks, and even once a live scorpion.

I did warn her on our wedding day that life might not be conventional and regimented, but that it would always be an adventure.

10

GRIMMEST

IF I WAS to name the dirtiest, grimmest place we ever shot, I would probably call out the black swamps of Sumatra. Although to be honest, in this category there are some strong contenders.

The Northern Territory of Australia, for example, is tough swampland for sure, and the mosquitoes that plague those backwaters can drive the best of us insane. I remember once filming in the outback up there and having had a brutal day of foot slogging through deep mud and thorny bush from where we had started the journey. It would soon be dusk, so it was time to set up camp.

We were all bushed after twelve hours of operating in 100 per cent humidity and crazily high temperatures. The mosquitoes had been getting bad but our local support team and our Aboriginal guides weren't complaining, so we muscled on. But by sunset those guys had all left, having hiked out to a roadhead to bring in the vehicles that would take the rest of the crew back to our production base in a town an hour away. That just left Dave and me, and a bunch of overnight emergency gear, all alone – again.

We set up camp and stripped down to our underpants to wash in the dirty creek water, trying to cool off as much as get clean. Then suddenly, in a matter of minutes, it was like every mosquito that was buzzing around our heads multiplied a hundredfold. One moment we were chatting away as we sorted our gear out, the next the noise of mozzies buzzing was so loud we had to shout to be heard. The air literally became thick with clouds of them. It was an insane sight.

Now, if you ever feel too small to make a difference in your life, come and experience the Northern Territory at dusk. You are *never* too small to make a difference. Those brutes can systematically reduce the toughest of men and women to gibbering wrecks – and fast. Truly, when it comes to mosquitoes, never underestimate their power to ruin your day.

By morning, our bodies looked like we had been rolling across drawing pins. We were swollen, bleeding, scratching messes – and that was even with a tent to crawl into.

It was another moment of awareness for me, that as a survivor I have always been more of a 'traveller passing through' rather than any sort of 'conqueror' of the wild. Strip away the tech and tools and resources, and pretty quickly we're in a similar boat to our ancestors thousands of years ago. Yes, we can put some dung on the fire and get smoked out to help keep some of the mozzies at bay. Yes, we can rub mud on our skin to try and stop the brutes biting. But ultimately, in the wild, we are never the strongest, toughest or most resilient. That prize goes to the animals. Hardened, honed, and refined in their ability to endure, adapt and survive. Animals are always in a constant fight to stay alive. As a result, they're pretty damn good at it.

Me, on the other hand? Rookie.

That journey in the mosquito-filled Northern Territory was one I will never forget for several reasons. Not least, the fact that it was the first time I drank my own urine. An ignominious landmark moment in my life, but a moment all the same. Urine drinking we have talked about, but another reason that episode stands out was that it was also the first time I ever found myself with a live snake's head inside my mouth.

Generally, people don't seem to like snakes much. (I guess, historically speaking, it goes back to the story of Eve being tricked by the serpent in the Garden of Eden, which led to the fact that man and snake have never really been the best of friends.) Snakes still account for a lot of deaths in many of the places I find myself in. Our former *Man vs. Wild* producer, Steve Rankin, got bitten by a fer-de-lance viper in the Costa Rican jungle and almost lost his foot and his life, despite getting to a hospital within the hour and having anti-venom at hand. Some snakes can be truly deadly. The Indian krait, for example, or the Australian taipan can inject enough venom to kill over fifty adult men. (Although as Piers Morgan once pulled me up on: So, they only bite men?)

The vast majority of snakebites are a total act of self-defence from the snakes' point of view, and snakes most certainly aren't the villains that much of the general population believe them to be. Although, I was once actively pursued by a highly venomous water moccasin in the swamps of Louisiana that came straight at me, skating rapidly across the top of the water in that classic S-bend formation. Its mouth was wide open, fangs exposed, most definitely intent on doing me harm. I had simply been minding my own business in chest-deep swamp water when I saw this thing skimming across the water towards me, and it's only because I just managed to grab a stick and whack the thing that I am here to tell that story at all.

Of course, there are always going to be more aggressive exceptions, but not all snakes are killers. I do occasionally come across some serious snake-obsessed folk and I'm all for it – it's just not my bag. You won't find endless glass cages full of reptiles round at our home.

Anyway, that day in the Northern Territory, I was making my way down a small, remote creek. The water was getting up to my waist when I suddenly spotted a 5ft-long file snake come swimming past.

I reached out, grabbed it by its tail and pulled it back towards me.

There are many different ways to kill a snake. The fastest way, which I had heard about but had never actually seen done at this point, is to grab a snake by its tail and then very rapidly swing it over your head and smack it down like a whip on the ground. Now, grabbing a snake by its tail is dangerous because snakes have a habit of being able to turn very fast and bite you – hence why I always say control the head first – but if you do this whipping motion fast enough and the snake is long enough, then in theory there is so much centrifugal force acting on it that the snake shouldn't have the speed or strength to turn and bite you. This is also actually a very humane way of killing a snake, it just doesn't necessarily look like it. I did it once in the Chinese rainforest and it worked a treat.

Anyway, here I was in the swamps of Australia, and I was just about to dispatch the snake and put it in my pack for supper later when one of our Aboriginal guides piped up and had a suggestion. He told us that some Aboriginal women historically would kill snakes by grabbing their head (better than the tail, as long as you pin the head first) and then – and this was the clincher – inserting the whole snake's head, alive, into their own mouths and biting down on the snake's neck to break it.

We all looked at each other and thought exactly the same thing: *you've got to be kidding.*

But then, of course, some smart alec on the crew goes 'Actually, Bear, that would be pretty awesome to see done . . .' and before you know it we are all stood there on the bank, in the middle of nowhere, with me holding this thrashing snake in my hand while everyone debates if this technique will work or not in practice.

The Aboriginal guide added that he had never actually seen it done, but it was definite folklore and he was pretty certain it would work. None of this helped me much. I had all sorts of images of the snake biting the back of my throat and then everything going south very fast.

Still, I decided it was worth the effort, and that if it worked it should technically be a very fast, humane way to kill the snake. I tried not to over-analyse the process, grabbed the snake and went for it . . .

I carefully moved the head of the serpent towards my face. Even this first bit just felt wrong. Then I opened my mouth as WIDE as it would go and eased the head inside.

Once I was sure it was in, I carefully clamped down my teeth to feel the back of its head and then quickly bit down hard and shook my head for extra purchase. This was the bit I was most worried about, because I knew my bite would force the snake's own mouth open, and if mine hadn't done its job properly then the snake could try to strike me as I withdrew it. I went to pull it out of my mouth super fast, and then let it hang down. Its neck was broken and blood trickled from its head. It shook and writhed as dead snakes do when they are first killed. My heart was thumping, we were all pretty adrenalized, and I looked back at Simon and Dan, our two cameramen.

Everyone was silent.

I breathed deeply and walked out of frame with the words: 'Well, that's dinner sorted.'

It had become a kind of standard response by me to catching my next meal, an accidental catchphrase. Whether I had a dead lizard, iguana, snake or scorpion in my hand, all too often it would genuinely mean one thing: dinner.

As I turned to come back towards the team, I heard Simon say the awful words: 'Bear, I'm so sorry, I didn't hit record . . .'

I couldn't believe it. After all that?

'You've got to be kidding me, right?'

He shook his head. 'We haven't got it . . . I'm really sorry.'

He didn't tell me for two days that he was joking.

He had got it all – and it was gold . . .

11

SNAKEBITE SURVIVORS CLUB

I WILL NEVER forget the time in Alabama when I decided it would be a good idea to get bitten intentionally to show viewers how to deal with a snakebite. Part of me had always wanted to become a member of the Snakebite Survivors Club – yes, it exists, and I always thought it would be a fun club to be a part of.

When I spotted a harmless but pretty aggressive-looking water snake I thought it was the perfect chance, not only to film a cool sequence that plays out everyone's worst nightmare, but also to show some solid survival advice about dealing with the aftermath of a bite. Plus, of course, I would get to become a legit member of the Snakebite Survivors Club.

What wasn't there to like about the plan?

A lot, as it turned out.

I grabbed the snake, held it by the neck between my fingers and called the crew over.

'I've got a great idea,' I announced without a second thought.

Everyone agreed, and the camera started rolling. This will be one to watch, I could sense them thinking.

'On you, BG,' Dan announced.

And it was at this point that I began to doubt if this was such a good idea after all.

There is something very unnatural about intentionally letting bad stuff happen to you. And on a list of most people's fears, snakes are always going to be high. I've never been too scared of snakes, just very respectful, and I have never handled one without my heart rate being substantially elevated in the process. It's hardwired in us. Instinct saying this is potentially dangerous, so get it right.

But the whole process of actually letting yourself get bitten, even by a non-venomous species, just suddenly came flooding into my brain, and my bowels.

Bear, you idiot. What part of you thought this was a good idea?

I paused.

Come on. You've suggested it, committed to it, now just do it. It will be a great scene. A memorable one. Your kids will watch this on You-Tube in years to come. Come on, muscle up. Let's do this!

It was the best, most convincing self-pep talk I could muster at the end of an eighteen-hour day of swamp thrashing in 100 per cent heat and humidity. It kind of worked.

OK. Let's do this . . .

I lifted the snake up, gave it a whack across the head – just to make sure it would bite me with some conviction – and then let it slide through my hand . . .

The plan was that I would then grab it by the middle of the body, raise it up, and watch it swing back and bite my hand for all it was worth.

I grimaced, squinted my eyes, and waited for the strike and the inevitable pain.

And I waited. No sharp, stabbing pain. Yet.

After a few seconds, I looked at my arm and the snake.

Instead of a writhing mass of wriggling serpent, striking at the hand that held it, all I saw was this limp snake, swinging gently from my hand.

I couldn't figure out what had happened.

The forked tongue hanging out of its mouth was the giveaway. The serpent was out cold, unconscious. The slap I had given it, instead of making it mad and aggressive, seemed to have knocked it clean out.

As the crew noticed this as well, there were a lot of mumbles of 'Good one, BG ... that's going to be a BAFTA-winning scene ... you muppet!'

You can't win them all.

So I still hadn't become a member of that Snakebite Survivors Club. But that got solidly rectified in Borneo a year later.

We were filming in the jungle and I was settling down for the night on a small 6ft by 2ft platform of branches I had built 30ft up a tree. It was dusk; Dan and Pete Lee, our other hero sound engineer, were with me in the tree, and the improvised tree house was almost finished.

'Snake!'

Pete saw it first, as it wriggled out from under one of the branches I had moved and shot along another branch up towards the canopy above. I quickly lunged to grab it and just managed to get its tail before it was gone.

That would be dinner. Usual routine.

As I went to pull it back, it wrapped its head and body around the branch and held firm. It wouldn't move. I saw that it was a viper. I couldn't tell what sort, but either way I would need to be super careful with it. You only ever get it wrong once, and never is that more true than when handling deadly snakes in remote places with limited supplies and access to medical support. I needed to be careful. But still I held on. All these thoughts happening in under a second.

I gave it one final tug and suddenly it just pinged backwards, and like a homing magnetic leash it locked on to my finger and started repeatedly striking and biting me.

I carefully pulled the head away from the bite, being mindful not to rip my skin, and dispatched the snake (I can't remember if I bit the head off it or severed it with a knife, but either way the snake was dead). The crew were in shocked silence and I was shaking.

'You OK, buddy?' Dan dropped his camera down and leant over to take a look.

There were two rows of puncture wounds on my hand. That was a good sign. Venomous snakes will leave two clear puncture wounds where the fangs have gone in and envenomated. Rows of teeth marks are more common with non-venomous species. The snake was indeed a viper, just not a deadly one. Lucky. Once more.

The biggest danger with a non-venomous snakebite is infection, and I remember being told in jungle training: *Snakes don't brush their teeth*. In other words, if you get bitten and it is non-venomous you still have to treat the wound very carefully. That bite will be full of bacteria, and in the jungle it is infection that is the biggest killer. Rather than crocs and other predators, it's always the small annoying things that get people.

Fire ants, for example. I once saw a big fire ant that had bitten Matt, one of our team, on the end of his penis. It was the first time I witnessed a grown man cry tears of pain.

Anyway, we treated my snakebite, and later that night I ate the viper over a small fire in my camp.

As I settled down to sleep I smiled to myself that I was finally a member of the infamous Snakebite Survivors Club.

12

DIRTIEST

THE BLACK SWAMP in Sumatra was a memorable double episode shoot for lots of reasons – not least because we were going to a part of the world that some years earlier had been brutally hit by a tsunami that had claimed almost a quarter of a million lives. The tidal wave had destroyed huge swathes of once thriving coastal communities, and turned forest into swampland that in turn had become a place of total desolation.

The mangrove swamps of Sumatra were now inhabited by all sorts of diseases and nasties – not to mention the fact that the crocs had developed a taste for human corpses. It wasn't a place to go into lightly.

The first part of the trip was to shoot another desert island show, in the Pacific Ring of Fire. It started with one of our most iconic free-fall shots: hanging off the skids of a helicopter, 10,000ft above this small island amid a sea of coral and blue. I look down, then up, then hang with one hand, cross myself (as has become my custom before a jump), and drop off, keeping eye contact with the camera as I fall. The shot gets used to this day by Discovery and National Geographic across lots of their promos. It's a fun one. Our eldest son, Jesse, whom I have taught to skydive, is always badgering me to recreate it with him alongside. We will some time. But right now he is still meant to be in training and shouldn't be messing about on his jumps quite yet. Or so we tell Shara.

The island shoot was tough. I had made the error of deciding to do it in bare feet, thinking it would be nice sand. As it turned out the nice sand was actually razor-sharp coral and I essentially spent the whole episode trying to walk 'normally' on camera, despite the pain. I told you, it's always the little stuff that gets annoying in the wild.

Talking of helicopters and footwear, one of the other moments I remember, which sums up my life, and my wife, happened while filming in the North African desert.

We had the Moroccan Air Force assisting us, and they'd provided one of their desert-camouflaged, fully tooled-up choppers for our Saharan insertion.

We loaded up, took off and slowly climbed to 8,000ft as we flew ever deeper into the desert dunes. We got given the two-minute warning from the pilot, relayed to us via the Air Force 'loadies' in the back with us. Simon got into position, tethered to the airframe, and I started to climb out. Simon always liked me to do a straightforward stable jump from the door, which was always the best way of guaranteeing he could keep me nicely in frame. I always fought against this and tried to use these jumps to do something original and scary and fun. Inevitably I won these battles ultimately by just going with the fun option, regardless.

I clambered down on to the skids, fighting against the 60-knot headwind. Then I lowered myself awkwardly into a position where I was hanging with my legs over the skids, head down to earth. I knew the shot from the second chopper would look awesome, with a lone man hanging upside down from a military chopper 8,000ft above the vast expanse of the Sahara desert.

'Ready, set, see ya . . .'

I straightened my legs and dropped away into the abyss.

When the episode aired many months later, I was excited for Shara to see it.

'Honey, check this one out.'

She glanced over, distracted from what she was doing, and half watched.

'Like it? Pretty cool shot?'

'What on earth are those socks you are wearing?' she replied.

'Socks? What do you mean?'

She reached for the remote, rewound the moment and paused it on me hanging upside down.

'Flesh-coloured socks? They're awful, Bear.'

I have a vague memory of being out of socks after the previous desert shoot we had done and remember borrowing some from one of our local guides. I hadn't given it a second thought. Until now.

It's the genius of Shara. Always grounded. Never letting me take myself too seriously and always with good taste. She sees the real stuff and is rightly never too impressed with the heroics. It has been one of the key founding factors to our enduring love. She is fun, normal, unimpressible and so damned smart. And she is almost always right. Those socks were pretty terrible.

Anyway, after that barefoot Pacific island shoot, we were about to start the jungle episode. First up, we had our customary briefing from the local rangers in the area. The plan was to drop in by chopper and make our way inland to higher ground. We had no idea how slow it would be moving through the swamp at this stage.

The rangers briefed us on the state of the swamp and I always remember them telling us that if we had any cuts or grazes, it would be worth either waiting a day or so for them to heal, or really to make sure they were well covered up and protected. The stagnant, rotting water of the swampland was rife with disease, and if ever there was a place to get a bad infection in any open wound, it was here.

The helicopter we were using was an executive Dauphin jet. It was super smart and not really designed for adventure operations, but the chopper we had planned to use had crashed a few weeks before while the crew were out scouting locations. The pilot had put the heli down in a small clearing in the jungle, thinking the ground looked solid enough. It wasn't. The skids started to sink into the mud and the tail rotor struck the ground. The result was bad. Really bad. But everyone managed to get out of it in one piece, including the badly shaken pilot, who now faced a 16km hike out of the jungle in leather brogue city shoes. Dave did a great job managing this disaster, and everyone knew they had been super lucky to get out alive. The chopper was deemed unrecoverable. It had been an expensive, painful day at the office for that pilot.

Nowadays, we are even more mindful to make sure the choppers we use and the pilots we hire are top grade. Taking away Hollywood guests and sporting superstars adds another dimension to my job that we will talk about later, but in TV you can't mess around any longer with sub-par equipment, relying on your wits and good fortune. Those days are rightly over, but there was barely a season of *Man vs. Wild* that went by without us hearing of another production having either a catastrophic helicopter crash or a very narrow escape.

I have lost count of the many helicopter trips we took that left me cold inside and amazed to be alive.

I particularly remember a tiny single-engine, two-seater chopper I flew back to civilization in after we finished a desert shoot once in Morocco. We flew over a huge mountain range being buffeted by the worst and most violent desert winds I have ever encountered. It was like being in a bad horror movie coupled with an even worse roller-coaster. The pilot was ashen and sweat was pouring down his face as we were thrown around like rag dolls in this tiny toy machine, with a barrel of extra fuel tied to the skid beside me. I genuinely didn't think the machine would make it, and it became another occasion when I mentally prepared myself for the end.

Strangely I was at peace, just annoyed that my life was over when I was still relatively young.

13

NIPPLE TIME

BACK TO THE Sumatra swamps, prepping for our helicopter infiltration, we were all happy that we were physically 'fit to go' – with no cuts or grazes to worry anyone. The medic was satisfied – and we started to get ready.

When the Dauphin helicopter landed at our LZ (landing zone) it was crystal clear that this chopper wasn't normally used for search-and-rescue-type operations. It had gold trim on the footwells and the wheel nuts were mirror shiny. We were told it was normally exclusively used by the Indonesian president – quite how it had ended up here, acting as our infil chopper, I still have no idea.

The look of horror on the pilot's face when Dave started gaffer-taping rope protection over the gold trim and attaching karabiners to the seat strong points was a picture to see. But even though the chopper's facade was different to what we normally used, we all felt comfortable that it was more than capable of doing the job required. After all, we had no intention of landing in the swamp . . . not this time.

Special Forces regularly use a technique called fast roping as a means of rapid troop insertion. I had done it multiple times during my stint in the military and we had used the technique often on *Man vs. Wild*. It involves a thick hemp rope and a pair of heavy-duty leather gloves. The gloves need to be thick enough to stop the intense friction heat that builds up incredibly fast, but malleable enough to allow dexterity and grip in your fingers. And unlike with a rappel, where you're physically attached to the rope as you descend, with fast roping you aren't attached by anything except your iron grip.

To minimize the risk, we tend only to fast rope either over solid ground from a relatively benign height, or over water (or swamp), where we can increase the altitude and the margin for error, knowing that ultimately, if you fall, you are less likely to die.

Nowadays, when we fast rope on *Running Wild* with a Hollywood star, because invariably the guests are rookies, I will put them on a top rope as well, operated by one of our team from inside the chopper. At the start of a *Running Wild* journey, the stars are not only adrenalized and often legitimately scared, they are also total newbies, so we don't take any chances. *Back up the back-up.* It's a vital mantra, especially when dealing with people who aren't trained.

It has taken a little transitioning from the *Man vs. Wild*, seat-of-your-pants, life-in-your-own-hands type of attitude to a *Running Wild* mentality of double-checking and re-checking everything. Certainly for the first few shows where I took stars along with me, such as Will Ferrell and then Jake Gyllenhaal, the risk management was not as thorough as it is nowadays. But we got there.

Anyway, back then in Sumatra, we were firmly in the 'screw it, let's do it' category.

Dave finished rigging the chopper and we loaded up.

'Fifty feet, hover and hold. I'll be your comms. Let me retrieve the rope fully before we move off once Bear is down,' Dave reminded the smartly dressed Indonesian pilot, who was looking ever more out of place in his captain's hat and tie.

The pilot nodded nervously. It was also clear he hadn't flown doors off and hot for some time, if ever.

The rest of the crew were already in position in the swamps, ready to receive me. We would then begin the actual journey. It was H-hour.

'Let's go.'

The approach was all good. Dave guided the pilot to the remote swamp location, who then brought the chopper into a hover at 50ft. The rope went out and I peered down.

The swamp looked horrific: reeds, mud and water were being blown everywhere. I could just glimpse the heads of the crew among the debris, and I could smell the stinking swamp, even from 50ft away. I knew that the humidity down there would be horrendous – but there we go, if this job was either safe or sanitary then a million other people would be doing it.

I eased myself over the door frame, grabbed the thick rope, and started tentatively to edge my way into a stable position under the chopper. It is always the first few awkward moments, where you end up half holding the rope and half holding the door frame, that present the moment of greatest danger. And of course, this is the point at which you are at your most dangerous height.

Concentrate. Time to switch into hyper alert and hyper calm mode. It's a place I know well.

All was good, and I started to accelerate down the rope. With the assurance of a rare soft, muddy landing rather than the metal deck of some rooftop, I allowed my speed to really build up. I knew it would be a cool shot. Into the swamp . . . rapid style.

By three-quarters of the way down, I was really flying. So fast that I knew if I had to stop, I wouldn't be able to using only my hands. But that was fine. The soft, stinking swamp was waiting below.

At that point, somehow the rope just caught my shirt and ripped some buttons off, followed instantly by an intense burning pain as the rope brushed against my bare chest at high speed. I knew at once that would hurt, but before I could think beyond that, I splashed into the swamp like a proverbial sack of potatoes.

Once I'd wriggled myself on to a small tussock of mud and lifted myself out of the gloop for a few seconds, I looked down to see a tennis ball-sized mark of raw pink flesh just above my nipple. It didn't look pretty. Then it started to bleed hard.

The words of the rangers team started ringing in my ear: 'Any cuts or grazes, don't go into that hellhole of a stinking swamp, if you want to live long.'

Well, we were firmly in it now.

And so began yet another classic *Man vs. Wild* adventure.

14

SILENT KILLERS

MOMENTS AFTER LANDING in that swamp, I remember catching a glimpse of a huge monitor lizard in the muddy water beside me. I just managed to grab it by the tail and swing it violently against a tree. It was as much in self-defence as an effective method of swiftly dispatching it. The reptile was at least 4ft in length, and we all ate it that evening sat around my swamp camp. Fans of *Man vs. Wild* will often mention that monitor lizard moment to me. I guess it's something about big reptiles in murky water.

Behind the scenes, it was so often the little moments that stood out for me. Such as the sight of our executive producer, Steve Rankin, who had joined us for this particular trip away, wading through the water as it got ever deeper. I was ahead, sat in a clump of sawgrass, watching as the crew approached.

'Guys, watch out, it gets deeper over here,' I shouted back.

I then watched as Steve began to edge forward, carefully moving his tobacco and rollies from his thigh pocket to his trouser pocket. Then, as it got even deeper, from this trouser pocket to his chest pocket. Then from his chest pocket on to the top of his cowboy hat, and then eventually to his hand raised up as far as it would go above his head. Meanwhile, his radio would be dead, his phone knackered, his notebook gone, but at least his tobacco was safe. I love it. That's another great image I always carry with me. Up to his neck in crap, yet always smiling, and his arm outstretched above him holding his baccy. Priorities, eh? Genius.

Every day, in truth, was stacked full of little moments like this. As I look back, it is these little things that stand out strongest in my memories. People's quirky habits, worried looks, moments of quiet struggle, and then times of outright laughter. Like the way our cameraman Paul Mungeam (Mungo) always looks as if he has a huge grin on his face when he is concentrating behind his camera lens, or Dan always putting sun cream on his ever-growing bald patch and jokingly making sure no one can see him. It's the little things that I love those guys for.

Steve, though, was a legend of a man who helped steer and improve *Man vs. Wild* so much at a few critical junctures in that series' existence. And through him I learnt so much. (Although the show almost cost him his leg and his life, as it was Steve who was the guy who got bitten by that fer-de-lance in Costa Rica.) After *Man vs. Wild*, Steve then went on to great success in America. It couldn't have happened to a better man.

I also remember one particularly close call in a jungle river a few days later, once we had transitioned out of the swamp to higher ground. The river was in full flood and we were down to a small skeleton crew, for some reason I can't recall. Flooding jungle rivers are notoriously dangerous, as they are always swollen with dirty, brown water. This hides so many dangers, at a time when the rivers are brimming with dead wood and tree trunks. These, in turn, act like submerged hidden 'strainers', as they are known, when logs get jammed and the power of the water then pins objects against them. Get trapped against one of these and it's game over.

The rule is always pretty simple: don't get into a jungle river after a downpour. Silent killers.

For the record, I always maintain that the most dangerous things in the wild are white-water rivers in flood, crevasses and avalanches, and then saltwater crocs. All these things can kill you in the blink of an eye if you get complacent. And they are indiscriminate. In other words, if you get caught out, it doesn't matter how big and strong you think you are . . . you're going to lose. It's why I always say choose your battles in the wild, because half of survival is knowing when to fight, and the other half is knowing when to run.

The director was really keen for me to do something 'cool' in this flooding jungle river, for a great sequence of 'man caught in rapids'. I wasn't so sure. Woody, a former SAS guy we had recently brought on to our team (and a legend of a man, who was such a huge asset to us for many years), had a very clear opinion.

'Don't do it, Bear. Jungle rivers in flood, kill.'

Dave and I reckoned it would be fine – risky but doable. It ended up with us all arguing on the bank, in the rain.

It's another reason why nowadays we don't use many directors on most of our shows. We know what we are doing and what we need. Having someone without a lot of adventure experience, pushing the team to do things that aren't always safe ('chasing the shot', as we call it), added a conflicting dynamic to an existing, highly performing unit.

Our job requires us to be on our game all the time, and to be able to make unclouded, smart decisions, often under pressure and against the clock. Arguing in the rain isn't part of that. The solution was to get rid of the directors. I loved that. And it was another example of us, through experience, rewriting the rules of TV.

Now, if we are ever told that we *have* to have a director, after some spurious edict from the network or broadcaster, we simply allocate one of our crew to wear the title. It's a box-filling exercise. We still might have a producer with us, but rarely a director. It makes for much more dynamism, and definitely more fun – and in our world, that also means much better TV.

Anyway, I made the ultimate call. I wanted to stop us all arguing, and I chose to do the rapids.

Mistake one.

It would be a quick shot: jump in, film it, and the crew and I would be done in ten minutes. We didn't bother to scout ahead, to check the river downstream.

Mistake two.

We were tired, it was the end of the day.

Mistake three.

This had all the ingredients of a royal screw-up. And it all came from one bad call. But I wasn't to realize that until I rounded the bend in the river.

So often, big disasters consist of three or four small, inconsequential elements that, on their own, might not mean much. But put them all together and you have a real problem. Forget to recharge a battery, lose a glove, break your compass, and suddenly up a mountain you are in trouble.

I instantly saw the huge wall of rock ahead, where the entire power of the river was banking up into a frothing cauldron of white water.

The real danger with these situations is under-cutting, where the water over time will have eroded a submerged shelf, and if you get forced into that, there can be no coming out. The power of moving water is simply too strong.

I saw the situation in a heartbeat. I have learnt that big, life-endangering moments are like that. The skill to survive them, though, is the ability to keep calm, act fast and see clearly when it counts. Such moments can sometimes last a fraction of a second. And in that time you have to get it right.

I knew at once that if there was a shelf beneath that wall, I was in trouble.

I spun my body feet first to be able to fend myself off the rock, and then glanced back at the crew following behind in a raft. I saw that it was slower than me in the water, and that they had now dropped back 30ft. That was bad. The second little element to this unfolding disaster scenario.

I took a deep breath, and in another heartbeat I was dragged under.

I felt for the wall but it wasn't there. Just a void, which meant one thing. A shelf.

Third element.

And before I knew it, I was in real trouble. In I went.

The shelf wasn't that deep – if it had been I would never have lived – but it was still deep enough that I was totally submerged and totally pinned. All I could do was keep calm, reach out my hand and pray that it would be above the water when the raft came by.

It was. And as the raft got thrown against the wall above me, I felt Dave's fingers grab round my wrist and pull me out. It all happened in seconds.

I came bursting out of the shelf, pulled by the weight of the passing raft and Dave's grip. I grabbed on to the raft's safety loops with my other hand and rode out the rest of the rapids.

When we hit the bank downstream we all knew it had been a close call. Very close. If Dave had missed my hand, and I had drowned in that river, it would have been no one's fault but my own. Three small elements to create a disaster.

In short, though, I had been an idiot and had overruled a trusted safety expert. It was a sober reminder that in the wild you only get it wrong once. And Dave, thank you. There were so many times like this we shared – and I'll never forget.

As we hiked out along the river bank afterwards, I always remember Dan saying to me, 'Bear, don't forget, it's only TV. It ain't worth dying for.' Simple truth, but I needed to be reminded of it all too often.

15

VOLCANO RESCUE

NOWADAYS, I AM SO cautious about how we operate. I should have died many times in those early adventures, and when you get low on 'lives', you have to get smarter, wiser and more risk-averse . . . or one day you simply won't come home.

This is why, having made it out alive from so many close scrapes (often small, unseen, forgettable moments, but ones that could so easily have gone the wrong way), I finally consider myself a bit smarter. I like to say that my knowledge now is the sum of all my near-misses. And on that basis, I can call myself 'experienced'.

Although, having said that, we never as a team get complacent. Interestingly, as I write this, I have just heard that the cave we filmed in with the comedian Rob Brydon and the cliff face I rappelled with the actor Warwick Davis have both since collapsed. Both incidents would have killed us and the crew. But how do you plan against forces of nature like that, apart from minimize your exposure to the danger, move fast, listen to the local experts, and say your prayers? There are no guarantees in the wild. It's an unpredictable environment. Wild by name and nature. That's just the game we work in.

The last memory I have from Sumatra involved filming on top of an incredible volcano. It was horrendous weather and the pilot kept changing his mind about whether he reckoned he could get us there.

Even in our books it looked sketchy, as volcanic sulphur mixed with thick mist, high humidity and jungle rain to create visibility like the inside of a wet smoke machine.

Eventually we spotted a tiny clearing in the clouds above and went for it. The pilot just managed to get us through, half tipped us out on the ground, and then left in a hurry, just about making it back to the better visibility below. It had been touch and go, but we were in. The only thing that was crystal clear was that he wouldn't be coming back to get us.

We would need to do what we needed to do, and then it would be a long hike out. Again.

We raced around filming a few quick scenes, arse-burningly close to the volcanic vent, showing the viewer how to survive if stranded up there, and then turned tail and started down, covering the ground at pace in an effort to be off the volcano before dark.

As we started into the tree line below, we suddenly heard shouting.

'Help! Help!'

We detoured left towards the sound and soon came across a group of kids and a very scared-looking teacher.

When they saw us emerge, looking like a dirty, sweaty band of militia, with rope and cameras slung over the crew's shoulders, their look changed from relief to trepidation.

Who are these guys? I could sense the teacher thinking.

Just then, one of the little kids piped up with a huge smile on his face: 'Look! Man vs. Wild has come to rescue us. We're saved!'

As far as that young boy was concerned, we had literally jumped out of the TV and answered their call.

At your service. How can we help you?

It was a fun moment to be part of.

We gave them a few snack bars, pointed the teacher in the right direction on his compass, and then we disappeared as quickly as we'd arrived.

Happy kids.

I sometimes wonder what they told their classmates and family when they got home, and if anyone ever believed them.

16

EXPOSURE

TIME AND TIME again, the wild has reminded me that I am really not as good at survival as I maybe used to believe. I mean, when I first set out to make *Man vs. Wild* I had a confidence in my own survival skills that was, in retrospect, a bit naive. The truth was that, despite having taught combat survival within my Squadron as a 'blade' with 21 SAS, and having trained in a bunch of places around the world, my skills were actually pretty limited. I had the basics, and a few cool survival tricks, and a confidence that only youth can bring, but the more I saw, the more I travelled, and the more I suffered, the more I realized that I had a lot to learn.

On *Man vs. Wild* I was endlessly meeting local survival guides who would be helping our team prep a journey before I arrived, and invariably they were incredibly well-versed in the local conditions, indigenous techniques and some amazing local flora and fauna hacks. Each time I was amazed by their knowledge. And by my lack of it. This sort of set-up became routine and it was very humbling.

I guess over the years I became a 'jack of many trades' but in truth never a master of any. We always moved too fast across too many terrains and too many countries ever to really have time to become a true expert in any one arena. It didn't take long for me to realize this, and as time went on I became ever more aware of what to me were my glaring limitations.

As the show grew ever more popular, I started to struggle more and more with the identity of being this 'hero' survivor that the show often portrayed. The truth was that there were many better survivalists, better bushcraft guys, better climbers, better skydivers; hell, it seemed to be that everyone I met in our work world was simply better, full stop. Not to mention stronger, fitter and too often much better-looking. I couldn't help but feel it wasn't quite right that I was the one in front of the camera. And that made me feel a bit of a fraud.

The more of these 'heroes' I met as we travelled, the more I began to doubt myself. And when confidence goes, it can be hard to recover. When we were filming and some new director assigned to our crew was watching me do things, I started to worry whether I was doing it well – the fires, the shelters, the knots, all silly stuff. I kept thinking they would be picking holes in it, and wondering how I had ever got this job.

Self-doubt can be crippling. Whereas before I had operated with a total freedom and assurance, and hadn't given those sorts of fears a second thought, now I could feel that confidence slipping through my fingers fast. And as I was finding out, when we start to doubt ourselves and our abilities, everything gets harder.

The mark of a champion is never the absence of fear or doubt, but rather how you respond when doubt and fear comes a-calling. That's the clincher. Because if the stakes are high enough, trust me, the doubts and fears will come. What matters is how we respond.

I had always thought fame would make me more confident. It would cure the doubts. Silence the fears. But it doesn't. Fame just makes whatever you have bigger. And the truth is, then and now, I know I am not as strong or capable as that *Man vs. Wild* image or any of our current TV shows sometimes project.

Even today, I am all too aware of this gulf between the TV hero and the reality of my own abilities and my many fears. It is this truth that has made me much more introverted as a person over the years,

I guess because a little part of me worries that people will find out how ordinary I really am. Yes, I have had some training in my life, and yes I have got to experience some amazing adventures with a little help from my friends, but at heart I really am a pretty ordinary guy.

In fact, if I was as strong or capable as sometimes gets made out, I wouldn't have had to endure half of the narrow escapes I have survived. I would have seen them coming from miles out and would have blown them to smithereens with bolts of lightning from my backside. Sadly, that's not how life works.

The way you deal with fear and doubt is to confront them. Head-on. Never run. Running always makes stuff worse. Keep moving towards the difficult. Acknowledge and then shake off those doubts, and keep going. Head down, into the wind.

And when it comes to fears about what others might think of you . . . forget them. Don't flatter yourself that people really care that much. Most people, as a wise man once said, are too busy fighting their own battles to care much for yours. As for what some director 'might' think . . . who cares? Let them do their job (which no doubt they are worrying about too), and just get on with doing yours – as best you can.

It might not be perfect but it's progress, and it's an attitude that has got me through some tough times. Focus on the little things, moment to moment. Build it back from the ground up. Two steps forward, one step back. *Just NGU – Never Give Up.* That mantra has lasted to this day among our wider team. And with my family.

NGU.

Not giving up is just so key to reaching the good stuff in life. And as a tool for dealing with hard times, it is king.

I remember when I was almost paralysed in a free-fall accident while I was in the military, I became acutely aware of how different my life could have been. Whenever I felt I was struggling or making next to no progress in my recovery, I would remind myself of the many other soldiers who were much worse off than me. And there were a good few of those in that military rehab centre, all fighting their own personal battles with untold amounts of courage. Never giving up.

What that knock did give me all those years ago was a huge jolt of gratitude for life. I wasn't paralysed. I could walk. And I would climb again. That was the goal. As simple as that. I then managed to fight my way on to a Himalayan expedition to attempt to climb Mt Ama Dablam, which had once been described by Sir Edmund Hillary, the first man to climb Everest, as 'unclimbable'. I made it up. Just. And it made me stronger.

Then on to a military-led Everest attempt, where I also made it to the top. Just. It's always just, to be honest. This time on Everest four other climbers tragically lost their lives. It was a huge shakedown for me. How did I make it through that one? Tenacity and some luck. A key combination for anything meaningful in life.

As you are maybe beginning to figure out by now, I really could have – should have – died so many times. My point is, what this simply means nowadays is that I am so grateful for life. Every day. After all, I should be dead (or at least paralysed), remember? So, I figure why not live life boldly now . . . because it's all a bonus.

And that's how I really see myself: as a regular guy who got lucky, who loves life and loves adventure. Adventure has always been one of the few things in my life that I could do well, and you cling to the things you love and that you can do naturally. I have my late father to thank for that. As he used to say: 'Rain is good, and be tough when it matters.' I've loved the outdoors ever since.

I have always liked the pressure of having to deliver in the big moments, up against it, in gale-force winds, when the chips are down. I like those times. But I am under no illusion that I am any better or tougher than any of the many other adventurers out there today. I'm not. But I love it all the same. And if I can do all these adventures alongside my family or with great buddies, then I'm happy. And that's kind of it.

In those early *Man vs. Wild* days, I didn't have the perspective I have now. Back then I was just desperately trying to stay above water, to prove myself in my skills and at my job. I was simply trying to earn a living. It sounds strange, but it really was like that.

When Shara and I got married, I knew I had to provide for her somehow. And we really didn't have much when we started out. Shara had some savings from her grandmother that she sank into a deposit on a barge we bought on the Thames in London, and I had inherited several thousand pounds from my godfather when he died, but had stupidly wasted it on a friend's failed business venture. (Lesson learnt: if you want to earn a fortune . . . do it yourself.)

Then, when we both lost our fathers so suddenly, and in our first year of marriage, it made that realization of being the sole provider ever more powerful. Life felt like it was getting harder.

I told myself, *You've got to work hard and risk lots if you're going to make it work in such an unconventional job as this.* That's how I felt every day – that I had a small opportunity to make something happen with *Man vs. Wild*. Whether I could do it or not was the big unknown.

Bear in mind that at this stage there was no other person out there doing survival shows in this way. I didn't have a road map to follow, or anything to copy. There was the bushcraft legend Ray Mears, who did some BBC shows in the UK, but his programmes were much more conventional and less controversially dynamic. And they weren't global. This whole new concept of survival in extremes that we had come up with was working, but it always felt like we were teetering on the edge of failure.

Just one accident, one injury or one dip in ratings, and the ride would be over. Plus, of course, no one knew how long the show would last. All I could hold on to was a determination to give it my all.

So on one hand I knew I had to keep pushing, keep risking, keep innovating, keep earning my living and building a career in this unusual way, but I was also becoming ever more aware of my own survival limitations. As our shows became more audacious, I had that creeping sense of not being good enough. Not skilled enough, fit enough, talented or strong enough.

With age and hindsight I now know everyone has these fears. It's part of all of our journeys, if we are honest. And it's OK. None of us are perfect. I know that now, and it doesn't matter. What counts are solid friendships and being honest, and trying to do our best through the storms that life throws at us. What matters is being kind, being fair, being determined, being humble. Never letting go. These are the real things of value. In fact they are the only real currency in life.

But at this stage, on my own, I wasn't seeing it like this. Instead, this creeping doubt just grew in secret. And that is always dangerous. Because bad stuff grows in the dark.

17

HEAVY ANCHORS

IT'S FUNNY HOW life is like this. The more we learn, the more we become aware of our ignorance.

It would be no lie to say I had started *Man vs. Wild* with the confidence to describe myself as a survival expert. Nowadays I would wince at that, and would immediately ask for that description to be changed if I saw it written up somewhere. 'Enthusiast' is maybe a better word (although a couple of nights with the Northern Territory mozzies actually sapped even that).

Maybe 'survivalist' is better. Although, if I ever have to describe myself now I always feel more comfortable with 'adventurer'. Certainly not 'explorer', which always sounds a bit self-important. And any exploring I have done of uncharted places is really very limited (there have only been a couple of attempts on various expeditions over the years), and pales into insignificance compared to the explorers of ages past who truly knew the hardship, fear, loneliness and deprivation of stepping into the unknown.

The truth is, I have had it all (well, mostly all) comparatively easy. And I am big enough and old enough (and ugly enough, as a former Sergeant used to say) to admit my many failings and shortcomings. It's just the truth.

Anyway, whether I am an adventurer or a survivalist, or a 'Boy Scout who is going to get people killed' as a veteran bushcraft icon once called me when asked his opinion on 'this new guy on TV called Bear Grylls', I take it all with a pinch of salt. None of us should define ourselves just by our jobs or titles, or by how others perceive us. We are all so much more. They don't put job titles on headstones, they write things like 'loving father' and 'beloved wife', 'kind friend to many'. I've always liked that.

One thing I was surprised by when *Man vs. Wild* was really building and starting to do well was what strong opinions other adventurers would have about the show. And to be honest, they were rarely positive. There was one particularly vocal Canadian survivalist who seemed to have it in for me from the get-go. And to give him his due, he had some fair points.

Like the 'Boy Scout who is going to get people killed' label, there is often a touch of truth to the criticism of the naysayers. Firstly, yes, I was a Boy Scout and still am, incredibly proudly so; and potentially, if someone with no skills went and intentionally got themselves into trouble in the wild, and had no back-up and copied something they had seen me do and just about get away with, yes they could get themselves killed – in fact pretty easily.

As I have said before, good technical survival or bushcraft is often quite boring. Stay put, take no risks, get yourself ready for rescue. *Man vs. Wild* took a different slant. A more dramatized slant, yes, but also a desperate 'your life is on the line' stance. The techniques could work, and do work, and if you have no other option they could save lives, and have done.

On the other hand, go into the back country alone, try and rappel using old parachute lines, slip, fall, break an ankle and end up facing off with a grizzly, and things won't necessarily end well.

My advice was desperate, last-ditch effort stuff. It is what made the show so fun to make and hopefully to watch.

I won't go into the many other less kind labels I've been given. Regardless of the criticism, I am only too aware of the mistakes I have made in the past, on and off TV. A public life never lets you forget. Some of them are known and a few are not, and like all of us, I have some I hope no one ever finds out about. I am flawed – we all are – but on the whole I hope I'm not too much of a bad guy. And I hope our shows have been received in the spirit in which they are intended.

I am not one of these people who has no regrets. I do. Where I have been unkind or hurt those I love, intentionally or unintentionally, I will always carry regret. And I am always sorry for my many mistakes. But I do try to move on, to leave things in the past.

That's a key one for our boys as they grow up: *When you screw up, admit it, apologize, do what you can to make amends. Then let it go.*

Guilt can be a heavy anchor.

18

ALWAYS MOVING FORWARD

THE REASON I used to find criticism hard was that I didn't really understand where it was coming from. It always seemed so personal and angry. All we were trying to do was make an awesome, fun, empowering TV show. But the truth is, I was entirely unprepared and ill-equipped to deal with that negative feedback correctly. I didn't have the wisdom, support or experience to be able to see it for what it is. I took it to heart, and it hurt. It made me question the whole thing – and myself even more.

The legendary late Kobe Bryant once said: 'Nobody criticizes good, only great.' In other words, criticism is sometimes a sign that you are doing something right.

I didn't always see it like this, though, and in the early days I took stuff way too personally. My confidence had been starting to wobble anyway as I had begun to realize my many inadequacies in the field, so this only compounded the problem. In turn, my doubts started to grow even bigger.

It took me a while to learn how to cope with this doubt. Over the years I've come to accept it. It's all OK. Just hang on in there and keep moving forward. Try and choose positivity. Don't take yourself too seriously, keep doing your best, admit mistakes, forgive yourself, give yourself room to feel the emotion, but also to see it for what it is.

For me, my Christian faith has always helped here. Not in a

particularly churchy way – in fact, I have never been particularly religious, but I have always quietly held on to the beating heart of empowering spirituality: that in Christ we are loved, despite our mistakes and our failings; that we can be restored and strengthened from within; that we truly matter, but that we are not the centre of the universe.

These are all good things to learn.

Almost always, we learn these lessons the hard way, and often through failure. But that's OK. These experiences are about seeing the world for how it is, trying to love those around us and living with gratitude. Do all of the above, and the doubts and fears tend to shrink . . . and eventually they get drowned out by light.

Oh boy. Plenty of lessons learnt. Almost always through pain.

For all the bad stuff that I have had thrown at me in the press over the years, though, there has always been a steady stream of wonderful life-saving survival stories pouring in. And they continue to this day. People who have found themselves stranded or lost or in trouble in the wild, and somehow remembered something from those *Man vs. Wild* shows – and it saved their life. A snowboarder who fell through a frozen lake, a mum broken down in her car in the desert, a boy trapped in a forest fire. There have been so many of them now and I feel a huge sense of pride that someone, somewhere managed to save themselves.

Although, as Mungo so pertinently put it: 'Well, you're hardly going to hear from any that have died due to your advice.' Thanks, Mungo.

By the way, Mungo, one of our core cameramen, has been one of the true heroes behind so many of my TV series, and he has stayed one of my best friends to this day. He shot my first ever TV series for Channel 4 with the French Foreign Legion, and we have been glued at the hip ever since. Even now he rarely misses an episode of *Running Wild* or *You vs. Wild* if he can help it, despite the fact he has a young family. There have been so many times when his friendship and encouragement have helped me at a tough moment. Without him, all of this would have been a lot more difficult and a lot less fun. Thank you, brother.

Just as I've come to terms with the opinions others have of me, I'm now often asked what I think of the new faces emerging in the adventure TV space. Many of them are ex-military, often ex-Special Forces. Often they climb big mountains, then want to get into hosting adventure shows. It was inevitable that it would happen and part of me is amazed it took so long. There will always be competition in the adventure TV field and if young men or women see people living a life of adventure and earning a living doing it, then why shouldn't they aim for a piece of the action? I know I probably would.

These guys often seem to get asked what they think of me, and to their credit they usually respond kindly. I am really appreciative of that. In fact, to be honest, almost all the key TV adventure players who have emerged over the last few years are really good guys, with legit back stories, and many years of military service that make my few years as a junior Trooper pale into insignificance. They are invariably the real deal and they deserve all their success. Any comparison can only ever be treated as a compliment really. And because the adventure community is essentially so small, I will always encourage anyone in it, and help wherever I can.

Part of the vision for our TV and film studios, The Natural Studios, is actually to help grow and build media careers for many of these up-and-coming adventurers; to help them create their own programming that fits with their goals and vision rather than just swelling a network's profit margins. The goal is to help new TV adventure hosts avoid the battles I had to fight to win my TV freedom, and for them in turn to be able to control their own shows. But more of that later on.

So, yes, I will always encourage the new guns, and help them where I can. It's a much stronger way to approach life than being guarded and defensive. Always keep moving forward.

The other truth, and this is not written as any form of self-deprecation or false modesty, is that I genuinely believe almost all of these new TV adventurers on the block are far stronger, more skilled, fitter and more capable than I ever have been. I really mean that. In terms of competition, I'm not even going to compete, as I'm pretty certain I'd get beaten.

19

LOCAL HEROES

THERE'S NO DOUBT that one of the great privileges of my job is that I have got to spend time with so many amazing people, and in some very unique situations. Whether it is the President of America, the Prime Minister of India or an Oscar-winning actor, the wild is a great leveller. I love that.

But one day, when I am old and losing my memory, I suspect it will be the other sort of people I will remember best. Whether it is Slav the Impaler, or the incredible San Bushmen in Namibia and our porcupine hunt, or the Emberá jungle tribe in Panama with their saliva-fermented liquor.

One guy I will always remember is Shelby, who lived right in the middle of the swamps of Louisiana. And when I say in the swamps, I really mean *in* the swamps. And a more brilliant, eccentric, tough man you won't ever meet.

Filming in the Louisiana swamps is always tough. It's a tough place full of tough people. My first time there was just after we had been film-ing in Mexico's Baja Peninsula, where I had been stung by a bee trying to get some honey out of a wild hive. I had a bit of an anaphylactic reaction and in a heartbeat my whole face had blown up like a balloon. The crew thought it was hilarious, and my family later called it an improvement, but either way it made shooting the rest of that episode tricky as I could only prise open one eye to see the ground ahead.

By the time we landed in Louisiana I was pretty drained physically. But a few days hanging out with Shelby in his old swamp shack – which was essentially a trailer balanced on sticks – was a healthy reminder not to complain about a bit of fatigue, a couple of bee stings and a swollen eyelid. Something I need reminding of.

Likewise, once in China. We had narrowly survived a typhoon in dense jungle, a very long way from civilization. We all looked like drowned rats: filthy dirty and exhausted after having battled through the Hainan jungle for the best part of six days, non-stop.

It had been our first episode in China and it was a big one for us and for Discovery. In fact, Discovery were using this episode to launch the channel into China for the first time. It was a cool thing to be part of, and as time went on from here, I recognized more and more the significance of the introduction of *Man vs. Wild* to a Chinese audience.

Man vs. Wild was the only show on the channel that was approved by the Chinese government, who were nervous of allowing a US-owned network to access the huge audiences in China. *Man vs. Wild* was allowed to air because essentially it was neutral, non-Western and rooted in nature and adventure, and if we could launch with an episode filmed in China then it would above all showcase the beauty of the country, right off the bat.

That moment has proved so key to my future. Since then we have sold over twenty million copies of our *Mission Survival* books there, and *Mud, Sweat and Tears* was voted the most influential book in all of China in 2012. Those sorts of things don't happen unless there is a fortuitous meeting of many different factors, and again I am so aware of a few key strokes of good fortune in China at a time when the country was only just beginning to open up to Western media. But as they say, timing is everything.

All I knew at this stage was that this jungle was a gnarly place. When research comes back and essentially says no one ever goes there, it's a good sign it's off the beaten track. It was a hard slog, and I ate a lot of grim stuff that I had managed to catch, from bat carcasses to rat's brain. I had also managed to twist a testicle – a fate I really wouldn't wish on anyone.

I had leapt off the back of a truck I had been hitching a ride with and had landed in a deep watery ravine from where our journey began. All helicopters had long been grounded with this impending typhoon, so the leap from the back of the pig truck was the best we could come up with as an alternative. The problem was that as I leapt, I had also somehow twisted my groin and caught a ball in a really awkward position. Mother love. I hobbled my way through that journey like an old man, half doped up on some serious painkillers from our med pack.

When the storm hit, we had sheltered for quite a while in a cave, which we'd figured was about the safest place to be when in a jungle in a typhoon. It had been mind-blowing to witness actually – the whole jungle being shaken as if by some giant running his fingers through the trees, and dead branches falling like confetti all around.

One of the biggest killers in jungles, ironically, is deadfall landing on you from a height. Branches break all the time in high winds and as a result of the activities of monkeys, and when they fall to earth from 100ft or so, it can be lethal. It's why deadfall accounts for so many indigenous fatalities in rainforests. In fact, our survival expert Woody had his back and both shoulders broken once by deadfall while teaching in Borneo during his time as Chief Instructor for the SAS jungle school out there. That's why caves can be about the only safe place in a jungle storm.

By the time we had got out of there and begun an epic seventeen-hour road trip back to civilization and the nearest city hub, we were all bushed. The journey was unending with all the road damage.

At about 3 a.m., after all of our senses of humour were maybe at a low point, we stopped at a junction, and I vividly remember watching this small group of construction workers digging a drainage ditch to help alleviate some of the flooding. It was the middle of the night in a torrential downpour, and this old guy was chest-deep in mud, swinging a pickaxe to clear a road. We were miles and miles from the nearest village or any form of transport or shelter. The old guy looked straight at me as we passed, gave me a quick smile and flash of his teeth, then carried on digging. I can still see his face today.

Whenever I get a little precious about something not quite working out as planned, or if I am struggling with being away or being tired or whatever other boo-boo we all at times feel, I try to remember that guy digging that road. Smiling. Little reward, save a job well done.

The more I live, the more I notice the real heroes in life. Quietly getting on with stuff. Doing their job, doing their best. Rarely a murmur of complaint.

And certainly not swinging through jungles for a TV show.

20

FORESIGHTS AND NUT SACKS

WHEN IT CAME to swamp-man, Shelby, we soon realized that this guy was unique.

He had a laugh that was unlike anything I have ever heard. He would let out a loud, open-mouthed cackle, and would then raise his right foot, slap his thigh, stomp his boot down, spit some tobacco on the ground and go off mumbling, 'Let's go get 'em . . . ha!'

We instantly liked him. Some years later he actually went on to have an on-screen role in a Discovery series about swamp dwellers, but it didn't last long sadly. I think part of the problem was that they had to subtitle everything he said.

Shelby was to be our swamp guide and a 'safety consultant'. That last part proved to be interesting.

He felt for the bone handle of his .44 Magnum that was tucked down the front of his pants and spat some more. 'Ha! Saddle up, boys.'

We flew around the narrow creeks in his flat-bottomed swamp boat at breakneck speeds, and he would laugh out loud if we got raked by the branches overhead, or if he turned too tight and bounced off a sandbank. Once he drove straight at a tree and the boat came to a bone-shattering halt. 'What the hell, Shelby?' we all said as we turned and looked at him. But before he could answer a huge snapper turtle landed with a thump in the boat, having been dislodged from a branch above us.

'Ha! Got him!'

And with that we were off again, holding on for dear life.

He only had two speeds, stationary and flat out, and every few minutes, if we hadn't seen a huge alligator or catapulted a turtle out of some tree, Shelby would simply draw his Magnum from his pants and loose off a few rounds wildly into the swamp. Just for fun.

'Ha!' Spit.

Those days spent with Shelby were priceless, and despite the incessant mosquitoes and dangers, Shelby just loved his swamp. The place 'self-administered', he would say. Simply put: those who weren't tough enough to deal with the alligators, humidity, stench, snakes, mosquitoes and 50°C heat didn't last long.

Shelby's actual home was a rotting caravan on a small patch of dry ground some 2ft above the water level and surrounded by swamp as far as you could see. It was an amazing sight. Sometimes you can forget that the world is full of Shelbys and that they truly make it a better place.

One morning, Shelby was a little quieter than normal.

'You OK, champ?' I asked as we loaded gear and camera equipment into the boat.

'Goddamn tooth, killin' me.'

He showed Dave and me his open mouth. It only had about four teeth left in it. One was looking particularly black, with the gum around it swollen, bloody and angry.

'That needs a dentist, Shelby,' Dave said. 'Seriously, buddy, get that looked at properly.'

'Dentist – like hell,' he replied. 'I ain't goin' to no quack. Ha!' Spit.

Dave looked in his med pack and found a bottle of clove oil. It would help ease the pain.

'Dab a small bit of this on your tooth and it should help in the meantime,' Dave told him. 'But go easy on the stuff. It's strong. And, really, you need treatment against infection and the tooth removed.'

We carried on for the day and as we were packing up at the end, Dave asked Shelby for the bottle of clove oil back for his med kit.

'Goddamn, I drank it – ha!' Shelby replied, spitting on the ground again. 'Tooth has never felt better. Ha!' Spit.

Dave looked at me. Shelby must have drunk the best part of half a litre of raw clove oil. An amount that would have killed most mere mortals. But not Shelby.

'Hey, one thing, Shelby,' I added. 'Your Magnum . . . I noticed you don't have a foresight on the end of the barrel. How come?'

'Ha!' Shelby replied. 'Cos one time, someone came round snoopin' at my caravan and I went to pull out my Magnum from my pants to shoot the mother, and as I pulled the weapon out, the foresight caught on my nut sack and ripped it clean open.'

We stood there laughing at the story.

'So I filed the barrel down and I've never had a problem since. Ha!' Spit.

'Hey, Shelby,' Dave piped up from across the track. 'I used to have some clove oil you could have put on that . . .'

21

SHERIFF IN TOWN

THE OTHER THING that happened in Louisiana was being sworn in as a Deputy Sheriff – and that was always going to be a cool moment.

We had a security team from the local Sheriff's office attached to our shoot, as word had got out that we were filming in the area. The show was really beginning to build in profile and popularity, and this was the first time the production team had deemed security necessary. For me, back then, the whole notion of having security where we were based was all quite alien and felt a little dramatic.

One of the deputies assigned to us was also a moonlighting stuntman in Hollywood. Jeff was a great guy as well as a true 'swampie', and after a week of wrestling alligators and catching catfish together, we had become good friends. Moments where you almost lose an arm or finger with someone are always bonding.

He told me that his boss was a fan of the show and that they had recently deputized Steven Seagal and prior to that John Belushi. Both icons and heroes of mine. When the Sheriff offered to do the same for me, it was impossible to refuse.

It would mean, on paper, that I would become part of their Disaster Response Team or DRT, joining their Louisiana parish as an official Deputy Sheriff. They made it clear that my military service would circumnavigate the need for any further training and that they had watched enough *Man vs. Wilds* to know I would be of value to their district in a disaster scenario. That was good enough for them.

I liked their rule-bending spirit already. I was in.

Without any further questions, they said I just needed to find time to pop down to the Sheriff's office to get sworn in.

That was where it got tricky.

This was one of the shoots where Discovery had sent an executive producer down to see the team in action. It meant that our team would need to be on its best behaviour. In short, it was annoying. Diverse TV, our producers at the time – and really, at that stage, our effective bosses – then got wind of this whole Sheriff saga.

No way, they said, could I go, midway through a shoot, and do this. No way.

The exec reminded me that we had gone through a flurry of negative press a few months earlier, all based around the show being 'produced' in parts. This was always bound to happen at some point, as the show had hit the mainstream by now, and it meant that the network and production company were paranoid about making sure there was nothing done on shoots that could in any way undermine the authenticity of the show. Sneaking off in the evening in one of the crew swamp boats to go drinking with local law enforcement and being sworn in as a semi-legit Deputy Sheriff was firmly in that category.

The producer came to see me and said in very plain English: 'If you do this, Bear, then it's the end of the road for your career. I will see to that. This just won't happen on my watch.' I saw this as a total overreaction and I felt he didn't really care about the show, that he was more worried about covering his own arse. I thought he was being ridiculous. After all, I would be out of the swamp and back from the Sheriff's office in under two hours.

I tried to reason that during those two hours I would simply be sat in the swamp by my campfire, just Dave and me, twiddling our thumbs. There would be no cameras with me anyway by that stage of the evening, as the crew would have gone back to their production base at a local hotel by then.

And the promise of a drink with the other deputies after the swearing-in ceremony was kind of tempting.

Both arguments fell on deaf ears.

The producer maintained his stance and reinforced the threat. No rule-breaking on his watch.

'Do this, Bear, and it's the end of the road for your career.'

I hope by this stage of the book I don't even need to tell you the rest of this story. And it all worked out sweet: we 'borrowed' a boat, made it to where Jeff was waiting for us at the roadhead, went to see the Sheriff, got sworn in, met the deputies, sank a drink or two, and were back before our fire had died out.

Who dares wins. And I am proud to carry my Deputy Sheriff badge to this day.

As for that producer, well, he left the show soon after.

And as for my Law Enforcement Officer (LEO) 'badge', I've produced it on a few occasions over the years, when I've been pulled over by the cops. It always raises a few smiles.

Interestingly, a few years later I was in a private airport terminal in Dallas when a Secret Service team entered the building. They were due to greet some foreign dignitary off a plane, and while waiting all together, we got chatting. After sharing a few stories and coffees, I jokingly showed them my Deputy Sheriff ID and asked them to check out how legit it really was.

They tapped away on their computer and up it came. Bright as day: Deputy Sheriff Bear Grylls, the State of Louisiana.

22

END OF THE BEGINNING

MAYBE THE BEST way of putting *Man vs. Wild* to bed is simply to give you a little insight into just one more of those early adventures. And maybe the simplest, best place to leave it is exactly where it ended, the last ever episode we shot of that original show, in Red Rock Country, Utah.

I choose this one not because it had the most iconic setting or action or stunts, but because it was the end. And for the first time I was just so acutely aware of how darned well this team worked. Seamless, smart, safe, fun. It just all functioned. Honed and refined through many hardships, failures and successes, and many trials and errors. Simply put, we all knew exactly what we were doing.

My doubts were now long gone because by this point we only filmed with directors and crew that I loved and trusted like brothers. We all knew our roles, and the goal was always clear: we were making innovative, empowering TV that was safe to produce, practical to follow and fun to watch. And through this we'd bring survival and adventure to a billion people around the world, every week.

Easy. And yes, we were now pretty good at it.

If I am ever travelling nowadays, filming our other shows, and I see some old episodes of *Man vs. Wild* on the box, in whatever remote country we are visiting, it is amazing to see the progression.

Season one, we hadn't got a clue. It was all one giant experiment.

I vividly remember the producer of the first ever episodes calling me when I was at the airport to fly out for our first jungle journey in Costa Rica.

'Bear, it's kind of working, but you've got to do a few things much better.'

'Tell me,' I replied, looking in my backpack for a pen and paper to take notes as I went up the escalator at the airport.

'Stop looking left and right when you are talking to the camera, stop being so breathy, stop wiping your nose every two seconds, and lower your pitch – you sound castrated for most of the journey.'

'OK, thanks for the honest feedback.' And with that I flew off to Central America.

Looking back, he was right. I was not good at this TV stuff initially.

I watched a clip of our first mountain show recently and there I was stripping down after falling through the ice of a frozen lake. I was dressed in a smart, striped cotton shirt and looked like I had just walked out of some office. I remember at the time thinking that I wanted it to look like I was just a regular person caught out in a blizzard, but the effect was totally the opposite. I just looked an amateur. Which I guess I really was at the beginning.

I actually feel I only started to get OK at doing the job after about season four. That urban episode in the Polish dockyard was a clear marker where I felt I was finally getting it. I knew our rules well enough to be able to bend them and make it all look great. And by Red Rock and our final ever *Man vs. Wild*, I also had the self-belief that I was finally getting good at this.

I now had the confidence to keep the bushcraft bits rough and ready – even more so now that the crew had shrunk to CTO – Core Team Only: a couple of my mountain safety buddies to help rope the crew into positions, a fixer, story producer, and a two-man camera/ sound filming team. I felt so much more relaxed.

CTO.

We had now learnt that slick was actually bad. Instead, we championed the unpolished. Mud on the lens? A few stumbles on words, or a trip over a log? Keep it – it's real. We knew this ethos and as a team we loved the spirit.

We knew what we were doing: we just dropped in and we went. Fast and fluid. We laughed when things went wrong, we took the mick out of each other. And we always had each other's back. For a crew that operated in some of the most hostile environments on Earth, I hadn't felt in such a 'safe' place for a long time.

It's that same dynamic that high-performing soldiers or climbers will know. Your environment might be deadly, but the trust and strength among the team makes it a stable and safe space to be. It's a great irony. And right now, at the end of *Man vs. Wild*, we were right there.

I would miss that, but I also knew it was time to end this particular adventure, and to start afresh.

If you stay on the summit too long you die, remember.

23

SMOKING TROUSERS

THE FINAL *MAN VS. WILD* shoot started with some classic 'improvise, adapt and overcome'.

The old WWII Dakota plane was fresh in from being used on the latest James Bond shoot. The pilots were great. Few rules, no bureaucracy, and willing to push it. And if you're good enough at your job, you can do this. The game is only working with people who are the best. These guys were.

The weather had blown us out and ruled out a para drop. We had to come up with another method of on-camera infiltration. A quick chat among ourselves, and Dave suggested we could try a 'touch and go' in the Dakota. While it was hurtling down the dusty runway, I could throw a rope out the door, lower myself out and roll away – avoiding the wheels – just before the plane took off again.

We all loved it. Original, fun and doable. It would look great.

A lot of the time on *Man vs. Wild* we had to improvise a cool adventure scene. No plan survives first contact with the enemy, as they say, and we were always having to go to plan B or C, due to weather, timings, permissions, you name it. It meant that we became very good at improvising.

We also became good at doing what we called 'dynamic risk assessments', to make sure we were always operating within our insurances. What this meant in practice was that we would discuss the plan and talk it through on camera, or at worst on audio, analysing the risks, the mitigations and why the plan stood up to scrutiny. It was essentially a spoken, on-the-fly, in-the-moment risk assessment, recorded in case it all went wrong.

We discussed the 'touch, drop and roll' plan, as we called it, with the pilots and then set to work.

When it came to the moment, the pilots definitely had more speed than we had provisionally asked for, and before I knew it I was being flung around like a rag doll, holding on for dear life to the end of the rope in a cloud of dust, dirt and friction burns.

At that speed, and right in the prop wash, I was totally browned out to any visibility I had hoped to have in order to let go at the right moment, to avoid the wheels and time it just right before the plane took off again. It was another of those moments where it all happens very fast, and you've got to stay calm and adapt the plan on the hoof. All while having your arse burnt to a cinder.

In the end, all I could do was let go and pray, as I got spat out at high speed in a heap of sand and grit. It made for a great start to the show, despite having two butt-cheek-sized holes burnt into my trousers.

It was spectacular country where we filmed that final episode. But then again, it almost always was. One of the many great privileges of my life has been to go to the world's most extraordinary places. That was certainly the case with *Man vs. Wild*, probably more than any other show since. Every week it was another amazing landscape, and another jaw-dropping wilderness. It was hard at times not to get blasé. I really tried not to.

Red Rock Country, Utah, was big, gnarly, hot, dusty, steep terrain. I always loved going there. It felt like a place rich in history, too. I liked that.

I was up and down cliffs, as always, and near the end of the journey I was following a river when I came to one particularly sketchy

Above: Stay low, watch your footing, follow me.

Below: The wild can sometimes beat the hell out of you, and sometimes it can simply make you smile . . .

Above: *Man vs. Wild* took me to the edge on so many levels. But it will always be the show that launched my career.

Right: No one ever said survival was easy . . . but it's always an adventure. Here battling a boa constrictor for Netflix's first ever interactive adventure movie: *Animals On The Loose*.

Below: The spiritual side of the wild is something that is hard to articulate, but it is about a connection to something greater and more powerful than just us.

Above left: This shot of me leading Hollywood icon Channing Tatum in Yosemite sums up a lot of the spirit of *Running Wild*. Steep terrain, crazy landscapes, buddy-buddy, life on the line, and never give up.

Above right: Pop icon Nick Jonas surviving the hard way alongside me. He never faltered and he threw himself into everything 100 per cent. The wild always rewards that.

Above: Always the best. The iconic Hollywood superstar Julia Roberts getting some BG rough treatment on our NBC *Running Wild* Red Nose Day Special in Africa.

Left: The stars always end up down and dirty on these journeys . . . but despite the odd protestation, I know they love it.

Top: Balance of a bird. Roger Federer nailing it as we descended into a deep gorge in the Swiss Alps.

Above: I was 7–4 up in the match of my life. Then in a heartbeat it was 9–9. From there, I got the yips and lost. The defeat still hurts to this day.

Right: With the heavyweight world champ Anthony Joshua for one of the ITV *Running Wild* episodes in the UK. Such a fun, humble guy.

Above left: In the desert with man-mountain and basketball legend Yao Ming for one of the Chinese versions of *Running Wild*.

Above right: On *Running Wild* with the incredible Shaquille O'Neal. We couldn't find any gloves that fitted, so we bought him some oven mitts instead. Improvise. Adapt. Overcome!

Below: Filming with Indian superstar Ranveer Singh. I love this shot as it shows so much: our incredible crew, the spirit of adventure and how the wild unites and bonds.

Top: Alongside our core crew we always hire a few local guides in each new location to help out. Local knowledge of the terrain has saved us on so many occasions. And, yes, it was often this fun.

Above: Our legendary safety crew, who have been beside me for so many years. Here paragliding out in the Alps all together: Stani, Scott, Meg, Ross and me.

Left: Banter with the crew is an ever present part of the job. And I love it.

Above: Del and me in the Fijian jungle. A friendship that was always destined to be.

Left: One of my favourite photos that sums up our journey. Del, trusting and fearless; and me diving out after. Always got your back, buddy.

Below: To have a trusted, humble, brave brother is a beautiful thing. Here in the mountains with Rupert, our CEO, planning, strategizing. As ever.

Above: My first time meeting a serving president of America. On a river bank in the middle of the Alaskan wilderness. Nervous but ready.

Below: The wild always brings people together in a way that everyday life rarely does. I love this photo. Just two guys laughing at silly stuff. I think in this instance, my bad cooking.

section of rapids. You'd have thought I would have learnt my lesson with white water after Sumatra, surely? I have now. Again, mainly through experience and near-misses.

I was using my rope to lower myself past a huge rock wall and around a bend in the river, through some fast-flowing water, to the bank on the other side. Normally it's a bad idea to mix ropes and rapids, as the potential for a snag-up is so high. (If you do use a rope in fast-flowing water, then never tie into it, as if the water pressure becomes too much you have no way off the line. I tend to take a large loop over one shoulder, so I can disengage easily if needed.)

I tied the line to a bunch of branches in the shallows and eased myself out into the rapids, heading around this rock wall that was blocking my path. If all was good beyond the rock wall downstream, then I would tug on the rope and signal the crew to follow me.

This sort of crew interaction had simply, and naturally, become a core part of our filming DNA by this stage, and I loved it. Take the viewer on this journey through the eyes (and often also the hands) of the cameraman. It made the journey feel even more visceral and real, and I loved being able to create a bond with the viewer simply by doing stuff with Simon or Mungo or Dan, as I would naturally on these trips. It made me feel more like we were in this together, which as far as the crew were concerned we always were.

The rope curled around the bend in the rock wall but there was a section ahead where I would have to slip under the rock shelf, beneath the water, while still lowering myself along the rope. It had potential for either a great shortcut or a monumental screw-up. I knew the chances of it being longer and deeper than the rope were slim. I had developed a good feel for this sort of thing, and from where we had recced the rock from above, the maths had worked out in terms of how far the underwater section of this traverse would go.

I took a deep breath, nodded at Dan behind his camera, and slipped beneath the water.

I popped up in the slack water in a perfect little eddy in the river. The white water was thumping around me, but the rope had guided me perfectly under the rock shelf and into this calm spot. It was one of the rare times that the plan had been inch-perfect.

I was about to tug on the rope to signal the crew to join me when I stopped. Just a few seconds. Let the suspense build.

Ten seconds later, I heard the first shouts from the team.

'Bear, you good?'

I smiled and let the silence – and the rope – hang slack.

I saw them tugging at the rope to see if I was still on it. Eventually, I gave it a tug back and the team appeared one by one around the rock wall and out of the rapids. It made for a great scene. And it had all been done safely.

Job done. Moving on.

24

LOOKS LIKE YOU'VE
BEEN IN A WAR

THE LAST FEW miles of that final *Man vs. Wild* journey was a hike through a desert canyon on the lookout for a local cowboy that our local fixer had found and hired to meet us at a certain location. As ever, we were excited to reach the end. It had been a long, hot shoot and the prospect of a few cool drinks and a good sleep was a fine one.

The first sign of life was horse manure. Fitting, I felt, for us. Often in the crap and always filthy dirty. And God knows I had sifted through enough animal faeces over the years. I had eaten berries out of the faecal matter of a bear in Transylvania, used dassie (rodent) droppings as a tea bag in Namibia and squeezed out and drunk plenty of fluid from the fresh dung of elephants over the years. To name but a few.

Generally, faeces can tell the survivor a lot. From what animals are in an area to how recently they passed that way to what they are eating. As well as their general health. I had grown accustomed to picking up the brown stuff and figuring out something good to use it for. Hell, I had even made fires from it and used it to smear on my traps to mask my scent. Not to mention encounters with my own, such as the time I was hanging off a rock face in Namibia and had felt some diarrhoea coming on. I'd eaten a raw snake earlier that day, having bitten its head off and drunk the blood, which I don't think had helped. But all of a sudden, here I was hanging some 90ft up a sheer rock face, inching my way towards the summit, when nature sent her wet warning.

I had hoped the diarrhoea might have waited until I reached the top. But diarrhoea isn't like that sadly, and when it comes, it comes.

'Turn the camera off, Si,' I shouted across to him, suspended on his rope from the top and filming across the dry face of the sheer waterfall. 'This isn't going to wait.'

I tried to get a secure footing and a solid one-handed hold, then quickly loosened my trousers with the other. It all happened kind of quickly and I managed to squirt brown water off the face and into free air below. The crew at the bottom were delighted. But it was soon all over, and at least it wouldn't be on film.

I got my trousers up, changed my hand grip to the other arm to shake some blood back into my limb, and then looked across at Simon.

The camera was still blinking away and filming. And he had a huge grin plastered across his face.

'Is nothing sacred on these shoots?' I shouted.

'It's gold – people will love it,' he replied.

I can't even remember if that scene made it into the final edit of that Namibia episode.

Anyway, back to the horse manure at my feet in Red Rock Country, and around here that meant cowboys.

I picked up the still-moist dung and announced excitedly that this meant we were getting close. 'Come on, guys, let's pick up the pace a little.' I flicked the remains of the manure from my fingers out towards the crew. A part of it landed beautifully on the lens. Bullseye. Simon hated it when I did that.

We pushed on. I didn't know what to feel. I knew deep inside that this whole *Man vs. Wild* journey was coming to its end.

The words 'pick up the pace a little' felt appropriate. It is how we had lived for the last seven years, filming six seasons of this show, with so many incredible adventures all over the world. Non-stop, endless enthusiasm and always at a relentless pace. Even if that pace sometimes meant making mistakes, incurring injuries and encountering failure. But then again, they say that the only thing worse than a bad decision is no decision. Don't pontificate for too long.

Generally it worked out.

My narrow escapes on *Man vs. Wild* taught me so much about life, adventure and survival that I could never have learnt from a book. As Lord Baden-Powell, the founder of Scouting, once said: 'A week in the field is worth a year in the classroom.' So true.

Man vs. Wild also stood for fun, which was always at the heart of so much of what we did. Not being afraid to try stuff that was at times ridiculous, and rarely made the cut, but that was worth the effort and at worst had made us laugh.

As a small team of adventure warriors, we had developed a great rule-bending, convention-busting, relentless spirit of positivity, even in the face of some overwhelming odds. Much of this has now become firmly written into our brand DNA and the shows that have followed. I am proud of that. We figured it out through doing it. Work hard, play hard, don't take yourself or the situation too seriously, know when to deliver and to focus, keep practical jokes going at all times, empower the viewer, and never give up. It was a good blueprint for life.

I hadn't expected to feel so much emotion as I hiked that last mile towards our extraction point. For so long the distances had always felt so far, those long hikes in and out of so many difficult locations just to start or end the filming itself. So many unseen miles. But suddenly this last mile felt too short.

Come on, Bear. Finish strong.

We finally spotted the cowboy on the top of the ridge and made our way up to him. Part of me wanted to hug him and part of me wanted to cry. I was still the only one on the crew who knew that *Man vs. Wild* was about to be over.

The cowboy shifted himself in his saddle and looked me up and down. Then he tilted his hat up.

'Looks like you've been in a war,' he observed in his classic American accent.

'Ah, some ripped pants and a scuffed jacket, but I'm good,' I replied. 'We had a pretty dusty infil at the start of the journey, that's all.'

'Well, as they say, it ain't how you start a journey, it's how you finish that matters.' He looked me over once more. 'Hell, you're upright and you're standing. That's good in my books.' He smiled, and passed me down his bottle of water to drink from.

And that just about summed *Man vs. Wild* up. Still standing. Still smiling. With so many of the crew still best friends to this day. That's something I am so proud of.

As Dave often used to say: 'We finish alive, we finish as friends, we finish successful. And always in that order.'

It's a great mantra.

25

GOING SOLO

ENDING *MAN VS. WILD* was always going to be hard. It was the show that launched my career. And anyway, why would you end a show when it is doing well? After all, in Hollywood they always say 'never quit a hit'.

But after six seasons of 'lone man battling the elements', I had a deep-rooted feeling that it was time to finish. We had survived so many scrapes and near-death moments that my gut was continually telling me to be smart and rein it in. The pace, the dangers, the near-misses and the time away from my young family were all catching up with me, and I knew it inside.

There are only so many close shaves that you can have before the odds catch up with you – and I had worked my way well over the quota, whether it be near-drownings in big jungle white-water rivers or close calls with crocs, parachutes, crevasses or rockfalls. It is indeed a numbers game, and when you are out there every week, year in, year out, it gets harder and harder to win every time – and in the wild you only lose once.

A producer told me early on, after only two or three seasons of *Man vs. Wild*, that essentially the problem is that this format can only really run for a year or so before either the audience gets tired of one man on screen, or that one man is dead. I've always remembered those words. And to be honest, I understood his point.

Still, I didn't have much else so we kept going. And going. Seven years later, I knew I had to break from it. That belief in always ending things five minutes too early than too late felt especially pertinent when it came to *Man vs. Wild*, and I was determined to finish the show on my terms, while the ratings were high.

The danger with TV networks is that they will bleed a show dry; even when the ratings start to dip, they will bleed it hard until it is wrung out, and they then discard it. I didn't want that to happen to *Man vs. Wild*, and I also didn't want to be simply a one-trick pony, known for one show that blazed bright, then died hard. I wanted to leave ahead of time, with the strength and ratings power to move on to even better and more innovative TV shows.

That decision took confidence – it is never easy walking away from something that is riding high. But I knew it was the right thing to do. Certain senior people at Discovery didn't like it much. In fact, they hated it. They told me, nicely at first, that I couldn't finish yet. I was still under contract for six more months. So I saw out another six months, then told them again. This time they got punchier.

I still wouldn't be permitted to quit, even if I was out of contract. The implication was that they owned me and I should just do as I was told. In truth, I knew that they would have pre-sold advertising on upcoming shows, and me quitting wasn't part of that plan.

I reminded them my contract had expired. They didn't care. That felt crazy to me. Plain wrong. I had done my time the hard way. I had to have the right to leave if I wanted to. So I pushed back and stood my ground. I was given another warning to return and start filming again, or else. When I held firm, they pulled the trigger, releasing a press statement saying they had cancelled the show and fired me.

Ouch.

They had played their first card.

I guess they were trying to make sure that if I was threatening to leave them, I wouldn't be able to work anywhere else. Then I would have no choice but to return, and they would be the good guys who kindly rehired me. I didn't like it at all. I knew it was simply a tactic to scare me into coming back.

It was now clear that they weren't going to roll over without a fight. But why should they just let me go? After all, they had a winning format and a high-rating show and, most importantly, advertising revenue against the next season. That was the problem. That meant I had to deliver more shows, or heads would roll. Mine included.

Their tactics made me worried, annoyed, scared and defiant, all at once. But my heart was saying something else: look up, you're not alone. Trust. Be calm. Be strong. Good will win. And above all, keep going. So that's exactly what I did.

At this time, I was also going through a very personal journey in my faith. Tough times do that to us. They challenge us to stand on what is solid rock not shifting sand. The things that really matter and will always be there. And part of that journey for me, as I faced this storm, was an overarching sense that I needed to trust the universe with all this.

These matters of the heart are super hard to articulate, but I have always had a deep sense of a presence in my life. A guiding hand. A force for good. And deep down, I couldn't hide from this feeling that I should hand over the reins of my life to that force, and stop trying to run my world myself.

Trying to run and control everything in your own life, especially when you can't see the road ahead, or don't know what's round the next corner, can be both futile and frightening. And not always that effective. Yet most of us are locked in this battle with ourselves, desperate to be the masters of our own universe. To me, that's always seemed to be the source of most of the world's ills. I've never really liked the whole idea of self-as-God.

Slowly, I was beginning to warm to this notion that if this

presence out there was real, and good, and wanted good things for me, and if this presence could see the road ahead nice and clearly, then maybe it would be a better guide through these storms. I could do the pedalling, but maybe I could give over control of the handlebars to a power that was better, smarter, wiser, stronger?

I tried it. Awkwardly at first, eyes shut, trying to articulate all this in my head. Handing over control of my life to whoever, whatever force maybe made this remarkable world, this universe . . . and me. Sat in the car, or sat on a plane, just for a few brief moments I tried to let go. It always felt so good. Actually, it felt exhilarating. And freeing. I started to do it a bit more.

I would begin each day on my knees, just like I used to see my grandfather do each morning beside his bed. It was a simple but powerful mental shift for me, away from trying to drive my own life, instead having the courage to relinquish control. I knew I was in a battle, with my career in the balance. I just tried to worry less about the outcome, and whenever I needed more energy, wisdom, faith or courage, it always seemed to come.

The more I did this, the more it actually took the pressure off me. My job was simple: have faith through the storm, be determined in my actions, smart in my decisions, trusting of my instinct, and free from worrying about whether I succeeded or failed. Whatever happened next was out of my hands.

This has been at the heart of how I have tried to live ever since.

I am not always successful in this endeavour, but it's my daily goal. And it is why still today I start my day on my knees. Know my place in the world, do my best, hitch my wagon to the higher power, and enjoy the ride.

26

FIGHTING FOR FREEDOM

AT THIS TIME, a great buddy of mine called Dave Segel, who had moved with his business and his family from the UK to Los Angeles, had quietly been helping and encouraging me to stand up against being 'owned' by any TV network in such a restrictive fashion.

Dave had always been a natural risk-taker and fearless in business, and he not only stood beside me and supported me with legal help, he also knew the LA media landscape back then much better than I did. I wanted to learn, and I knew that what he said about taking back control over my TV career would be a smart move long term. Whoever I was making shows for.

The problem was that no one had ever agreed a deal like this with Discovery Channel before. A few had tried, saying they would quit their shows unless they had more ownership and influence over production – but it had never ended well. Understandably, the senior team at Discovery weren't going to let what is known as 'talent' hold a global channel to ransom – and in turn set a very unhealthy precedent. And when you are a billion-dollar company, it can be pretty straightforward to win battles against individuals. Sadly.

Still, despite all this, in my heart I knew this would be key for my future in TV, if I really wanted longevity and a fair slice of the pie. Otherwise, all the spoils would simply keep going to the channel – and that inherently felt wrong. After all, it was never their arses getting bitten by the snakes.

Dave and I also knew that if down the line I wanted to do programming for any of the other big US networks, this battle for my rights and freedom was one we had to win.

The year all this was kicking off, I had been in LA filming a new series for Discovery called *Worst Case Scenario*. It was a really fun series to make, all urban-based, with more stunts and crazy stuff happening every day of the week than I could ever have imagined. I loved it. One day I would be driving a car at breakneck speed down a narrow mountain road with no brakes, simulating what to do if you ever experience brake failure, and the next I would be jumping out of an eight-storey building into an airbag, simulating surviving a gas explosion.

The family came out with me, we rented a great place on Malibu beach, and I went to work every day. It was good, well paid, but I still knew I had to confront the beast about ownership, rights and my TV freedom, if I was ever going to build a serious media future that wasn't weighing on my shoulders, not to mention being vulnerable to an abrupt ending if I got injured or the shows got canned.

Apart from the many narrow escapes with my life over the years, we had also narrowly survived another potentially catastrophic ending to *Man vs. Wild*, and probably any future TV career, when the press started running stories about a whole bunch of Discovery series being full of supposedly 'staged' scenes. At this point we were on about season four of the show and just finding our groove. But in TV terms we were still relatively young.

This all came at a time when Discovery Channel was about to float on the stock market. The senior guys knew they couldn't afford a bunch of negative press, so I heard that the directive had gone down the chain to get rid of any of the channel's shows that might contain any 'produced' elements that could be construed as faked.

On that basis, almost every TV show would have been axed. (It's worth noting that almost all TV has an element of pre-planning, mainly so that a one-hour episode can be shot in a week rather than a year of 'waiting' for things to happen.) At this realization, Discovery instead simply axed the weaker shows and invested more in the winning ones. It was a lottery really, but we got lucky and survived a major culling of shows at the channel. (Although Jane Root, the then head of Discovery, did play a huge, positive role in fighting for and protecting *Man vs. Wild* as a format she really believed in as a force for good at Discovery. I will always be grateful for that.)

The irony is that, from that pivotal moment, we really went on to establish and cement the show. We started to smash the ratings in America and, perhaps more importantly, in almost all of the crucial new international markets. It meant that the show became a solid hit worldwide for Discovery, in terms of ratings and advertising revenues.

This was why no one wanted some upstart Limey host threatening to quit the show. *Don't quit a hit, Bear . . . you might never get one again.* I heard these words over and over. But that's an attitude driven by fear not vision.

I knew this plaster had to be pulled off if I was ever going to grow some new tough skin underneath. During filming for *Worst Case Scenario*, the production assigned to me a young guy called Delbert Shoopman, as a driver and assistant. I could never have imagined how this baseball cap-wearing guy, being paid a few hundred bucks a week, at the bottom of the production's giant food chain, would go on to become one of the most important people in my life. But then life is like that sometimes.

On our first day as a family in LA, we had turned up at this house in Malibu that the production company had rented for us. But as soon as we went in we could see that it was infested with flies and stank of cat pee. It had been a huge decision for Shara to come out, with the boys being so young, and I just knew this house wasn't going to work, leaving her all day with flies and urine (that was my job, not hers). I pulled the plug straight away and said, 'Leave it with me, honey, I am going to sort us out somewhere nicer and better. Don't worry.'

Thirty minutes later we were all sat in a Starbucks drive-through, in the same pick-up truck Del had collected us from the airport in earlier, luggage still in the back, casting votes between the family over where we should go. Either we stay in the fly-infested place, or go to a motel until we find a new place, or stay in Dave Segel's in-laws' small, wooden beach house, which he had kindly offered us as a stop-gap. The casting vote came down to Delbert Shoopman.

Del looked a little nervous, but he went for it and we plumped for the wooden beach house. It was a great decision. From that moment on, Del was a firm part of our family and our young boys loved him. I don't think he knew what had hit him.

As a family, we tend to have pretty good judgement when it comes to people. And when it came to Del we knew within a few weeks that he was special. Kind, honest, loyal, hard-working and always punctual, we soon included Del in everything.

It was another moment where it felt like something had led us to this young, quirky and, on the face of it, pretty unlikely person. But experience has taught me that those types are always the best.

27

JOIN ME

SOME THINGS IN life have a sense of destiny, and meeting Del was definitely one of them. Very quickly it became apparent to me that Del was destined to be a lot more than a PA.

We drove together to locations every day, often at 4.30 a.m., followed by long days filming, then long LA traffic drives home. We got to know each other really well. He was a ball of talent and energy and potential, just waiting and working towards his own break.

The production company that employed Del had hired him to work literally at the bottom of the tree, and he was treated like it. I felt inherently protective of him, and it was obvious that he wasn't going to get the chance he deserved while stuck working for those guys. Then again, as Del says nowadays, you've got to go through some of that stuff to make sure you are in the right position when your opportunity comes along. You could argue that Del was exactly where he was meant to be at that time.

We became like brothers during those months, and when the show eventually wrapped I asked him if he would ever consider quitting his job and instead coming to work for me. Mission one would be to help me break away from all the agents and agencies I had found myself being repped by. I began to feel they were essentially hungry for the earnings from the show and that their true loyalty was just as much to the networks, rather than to me. I just didn't trust them, and their fancy offices and flash cars. It didn't smell right.

Money can so often cloud someone's judgement. I had always cared less for the money than for wanting to work on my own terms, on the projects I believed in, with the people I loved. The other motivating factor for me was to make sure that whatever I did also gave me as much time as possible at home with my young family. Agents don't always like that because it typically means you're bringing in less money. So we had some expensive, painful break-ups to complete.

I paid everything that every agent asked for. It was more than I ever owed them but I didn't care. I was going to rebuild this ship from scratch – bigger, better, smarter than ever before. And those guys wouldn't own a cent of it.

This process cleaned me out of all the savings I had diligently put aside over seven years of risking my life on *Man vs. Wild*. All of it. That part hurt. I really was now back at ground zero, only this time I had a young family to look after. But I still knew it was right. And to Shara's credit, she never questioned this decision. As ever, she encouraged me and she supported me. Although the legal letter FedExed from Discovery that Christmas shook her pretty badly. But with an amazing wife and family beside a man, he can do most things. We held firm.

I also had an incredible agent in London, a real Rottweiler of a man called Michael Foster, who had fought my corner so hard in the early days of *Man vs. Wild*. He had actually done a really solid job for me and had greatly increased my fees, but essentially the same problem remained: I was always going to be a gun for hire and had no security if anything bad happened. And in terms of the shows, the productions and the royalties . . . I owned nothing.

For all Michael's many strengths, he was also prone to some serious outbursts, and I had begun to spend a lot of time apologizing to people in my work world for my agent's attitude. I didn't like that. You always have to operate with respect for people, and to try and spread goodness and kindness amid the deals and negotiations. Those are the bits that people remember. I started to dislike some of the negativity that was being created, even if, overall, Michael was winning for me.

The LA agencies weren't much better; their anger was just more hidden. You might argue that's worse. Either way, I wanted a total clean slate. No agents, agencies or one-sided network relationships. We were going to do this a better way.

I said all this to Del. For better or worse, I was going solo. It would be highly risky, undoubtedly scary, with a very slim chance of success and with no previous precedents that said we could even achieve what I wanted. It would take a brave man to throw his job in and join this ship heading into battle with the Hollywood Armada.

I also couldn't promise much of a salary, and he would be working from the lobby of Dave Segel's company HQ in Santa Monica. It really wasn't much of an offer, truth be told. It must have looked like there was a lot of downside.

To his credit, Del never even hesitated. At worst, he knew we would have an epic adventure failing so boldly . . . but if we got lucky and managed to pull this off, it would change both our lives for ever.

On day one, with Dave's guidance, we set up the company BGV (Bear Grylls Ventures) . . . and a new adventure began.

28

BATTLE LINES

MEANWHILE, DEL, ME, Dave and his legal counsel whom he assigned to us – an amazing man called Todd Burns, a rare LA lawyer who was as smart and honest as he was anti-corporate – held firm and resisted Discovery pressure to go back and restart the next season of *Man vs. Wild*.

Part of me felt liberated saying no. But a big part of me was terrified too. The legal letters that were coming in were getting ever more aggressive and unpleasant. I kept feeling that I really should be more worried. But at the same time I felt an overarching sense of peace and assurance about our path. I kept trusting. Kept fighting. Kept listening to that inner guide. And I kept going.

We kept holding firm, until finally, thankfully, we found a positive resolution for all. It wasn't accomplished through legal wrangling. Instead, we went to see them face to face, explained my corner and listened to theirs. The great peacemaker through it all was David Zaslav, ironically the very top guy at Discovery, and one of the good ones. I always felt that it was the senior legal team and the head of international productions who were the ones out for my blood, but when I went to see them properly, and once the legal threat abated, we actually had some pretty fair and balanced conversations. And to give them their due, in the end they acted with real grace, and I was free.

Ultimately, I believe it was David who brought some calmness and fair play to the situation. For that I have always been grateful, and he is still a great friend to this day.

Man vs. Wild was over, and I was cut loose. That was the good part. The hard part was that I now had no TV show, and no savings to show for all those years of work. For the first time, my destiny was in my own hands, and I had the ability to go out, start again from scratch, and make sure we devised and created contracts moving forward that respected us as creators, originators, producers and hosts. And therein would be our value.

Now all we had to do was persuade a TV network that this was a great idea.

The parting words Michael Foster gave me before I embarked on this going solo journey were: 'Don't be an idiot. If you go to a big US network like NBC or Fox, they will force you into a much tighter and tougher contract than your current one with Discovery; they will be much more ruthless and less loyal than Discovery; and any dip in ratings and you'll get canned. In the US that then means you are soiled goods, and it will be almost impossible to launch a new show elsewhere.'

As to the notion of having multiple shows on multiple networks, which I had told him and my LA agents was my ultimate goal, they all just laughed. TV doesn't work like that. If you are on ITV then you can't also be on Channel 4; if you're on NBC you can't also do cable. And as for doing both Netflix and Amazon Prime at the same time . . . it could never happen.

I always questioned all this, and often wondered who had written these rules in the first place. I look back now and marvel at my naivety, but fortune always favours the brave, and I didn't take much notice of all those naysayers. I wanted to at least try.

I don't know if I would be so brave again today. It was a real roll of the dice, but backed up by two things. First, a belief that if a show was good enough and the ratings strong enough, then everything would be negotiable; and second, a conviction that this risky route was somehow the right one. Trusting my heart.

I was still starting every day on my knees, and I had never felt so bold in my life. Maybe totally unjustifiably, but anyway, screw it, I was going to give it a go . . . or die trying.

29

HUMBLE BEGINNINGS

AS FAR AS I was concerned, I now had a chance to build my own future. Plus I had found an awesome wing man in Del – albeit not exactly a conventional LA graduate hire.

If you were setting out to build a global production studio then logic would say hire on a killer CV and strong track record. Instead, I had hired Del, a country boy and former US Air Force grad who was earning less than a waiter. But I trusted him and sensed something special in him – plus he was ambitious and smart, and willing to go all in. That's a key quality in any champion.

The goal we set together was to be able to make shows that we owned, underpinned by the core values that we shared. For me, it was about being able to make shows on my own terms. The days of being dictated to about how, where, when and for how long I went filming were over. I had a young family and if I was going to do TV again then we were going to do it on our terms or not at all.

It felt very empowering. And I knew deep down we had done the right thing. It had been painful and scary, but I had held on, kept my eyes looking upwards, and we had prevailed. Now it was time to go and make use of that freedom.

Dave and Todd had done what they had set out to help me do. They had done so for no other reason than friendship, and that is rare in the world of TV. They helped Del and me establish BGV, lent us space in their office and then sent us out to fly. Del and I were now on our own.

The problem was that we essentially had nothing. No shows, no funds. Yet we were strangely excited.

The first show we got to make was called *Get Out Alive*, and it was my first time making a series for network TV in America, for the mighty NBC.

We had partnered with Electus productions, and a highly charged New Yorker called Chris Grant. The two of us had hit it off instantly. Chris is straight-talking, scarily bright and highly motivated – and he's as city-born and -bred as they come. Despite our different backgrounds, we made such a good team. We both knew our strengths. His included persuading one of the biggest, most powerful networks on Earth to back me – and for that I will always be grateful to him. He's a great buddy to this day.

Bear in mind that as far as the world was concerned I had just quit *Man vs. Wild* – or as Discovery made out in the press, I had been fired for 'creative differences', which really meant insubordination.

So, for us to pull off a huge-budget primetime American network series, straight off the bat, with no 'pilot show' or anything, was a huge coup for Del, Chris and me. It was also logistically a huge show to put together, taking a cast of regular American pairs (fathers, sons, best friends, mothers, daughters) into the wild mountains of New Zealand's South Island, and setting them some huge obstacles and challenges to overcome. All the contestants were rookies, and to add to that I now had a crew beside me of some 150 people, including four helicopters, tens of mountain guides and an entire village of a production team. It was a whole new beast, on every level. I was pretty nervous.

The saving grace was the fact that I had brought along almost the entire *Man vs. Wild* team to help film it. That, as ever, tends to make the difference between a great experience and a less enjoyable one. Producers think that a cameraman is a cameraman, or a sound man a sound man. The really great producers know that is far from the truth.

I have always believed that *you have to hire the best*. It pays back many times over. And *loyalty among friends really matters*. Two more mantras that I try to live by.

As for the contestants, they got the adventure of their lives, as episode by episode I whittled them down. And as the journeys got harder, and they were really put to the test, some beautiful relationships and back stories began to emerge.

I was looking for mountain qualities not Olympic abilities. That meant that good teamwork and quiet courage mattered more than muscle. It was an original take, but the part I loved most was seeing, and rewarding, qualities such as selflessness, humility and kindness just as much as I did the summit glory moments. It is a format of value-based programming that I have used across so many shows since, and I love it because it allows some real heroes to emerge and shine. You don't need muscle, you need heart.

By the end, we had truly had a blast. And I got to give a million dollars to such a special couple, a father-daughter combination who had shone bright, dug deep and never given up. Their resilience and their spirit carried them to victory and I wept unashamedly at the end.

Our new TV journey had begun.

30

RISK NOTHING, GAIN NOTHING

I WAS SO fortunate to have been given the backing of NBC to make such a big-format show. The man I owed that to was a bright and influential Brit running NBC factual TV called Paul Telegdy. It had ultimately been down to him that we had been granted the opportunity.

The problem was that the show didn't rate. It wasn't a disaster, but it wasn't the huge hit they had hoped for and invested in – by any means. Borderline ratings on network TV meant one thing: the show was canned.

Bad.

This had been our first BGV project, and it was what we had fought so hard for. We had battled to win the freedom to make TV as we thought best. But the truth was, it simply wasn't good enough.

Again, the words of my former agent, Michael Foster, warning me against doing network TV in America rang around my head: 'Deliver bad ratings and they will suck you in, chew you up and spit you out. Then you're screwed.'

But I didn't get too worried. It was always going to be hard to strike gold twice in a row, after *Man vs. Wild*. What mattered was keeping going, and to live the truth that failure simply makes us better. So we did just that: we kept going regardless, and never once considered that this new venture might not work.

In truth, it was the faith of Paul Telegdy at NBC, and his unwavering belief in the positive values through adventure that we stood for, that kept us going. He saw the future potential if we got the format right and gave us a precious second chance. As I told him as we left his office: 'We risk nothing, we gain nothing. Come on, let's go again. NGU, Paul.'

'NGU?' he replied.

'Never Give Up.' I smiled at him.

If truth be told, I think Paul backed me partly because I was a fellow Brit. It's a small club in LA, and I think he found it refreshing to have someone else gently taking the mick out of him and the whole LA scene. Not many Americans could do that with him. Whatever the reason, I owe him, big time. He gambled once more.

Paul had seen how successful a couple of our *Man vs. Wilds* had been when I had brought a Hollywood star along for the ride, first in the form of the legendary comedian Will Ferrell. The show had been a huge ratings hit for Discovery, and this hadn't gone unnoticed in Hollywood.

Discovery shouldn't be getting ratings like that. Those sorts of US numbers had never really happened before on cable TV. But as far as we were concerned, we'd simply included Will's journey with me as part of the new season of shows, and hadn't really thought too much else about it. We'd called it *Men vs. Wild with Bear Grylls and Will Ferrell*.

After that, we'd done one more, during that last season of *Man vs. Wild*. This time with the Hollywood heart-throb Jake Gyllenhaal. It was another winning episode in terms of viewer response. People had simply never seen such huge stars uncovered and raw like this. No entourage, no make-up, just out in the wild, battling against the elements. And not always winning. That was in itself pretty unique. And I loved it.

These were the pioneer shows for what eventually became *Running Wild*, and looking back, both those guys got the adventure they wanted in spades. We would never push our guests that hard nowadays, but back then it was much more a case of us simply filming a normal version of *Man vs. Wild*, but taking a rookie actor along with us. Albeit two of the most well-known stars on the planet.

Ironically, both journeys threw up some of the toughest conditions we had encountered for a long time. Will came all the way up to the Arctic to join me, and it almost killed him. It was minus 30°C all day and night, with deep snow, steep mountain terrain and only a Twinkie bar in terms of supplementary food. Only his undying humour and quiet fortitude got him through, and my respect for Will grows every time I catch a clip of it on air. He really is an amazing man to whom I owe just so much. Without his courage to come along on that first ever one-on-one celebrity adventure, the rest of my career would no doubt look very different.

As for Jake Gyllenhaal, we went to Iceland in winter. It was never going to be easy. But Jake is without doubt one of the gentlest, smartest guys you'll ever meet, and he's also got the adventure bug in spades. Just halfway through that trip we were caught in the worst winter storm in Iceland for over a decade. Hurricane-force winds high up on the glacier we were on meant that for the first time ever we were simply forced to stop filming and to start surviving for real. Jake just kept smiling and kept going.

At one point, I saw Dan Etheridge our cameraman suddenly get picked up by these hurricane winds and blown across the freezing tundra. He was still holding on to his camera as he flew like tumbleweed across the ice. I grabbed hold of Jake and we cowered against the wind. He was loving it. We all survived (we even managed to find Dan again when the wind finally dropped), and we had another ratings winner of an episode in the bag.

Paul knew that this new style of show we had stumbled upon worked well. It had already created some waves on cable and proved itself universally popular with both stars and audiences. If NBC was to gamble again and back us, then that would be the format.

So *Running Wild* was born – and it has never looked back.

31

RUNNING WILD BEGINS . . .

WE QUICKLY LEARNT on *Running Wild* that no Hollywood star wants to slog to Siberia to film twenty-hour days, dragging themselves across frozen lakes and up steep ice cliffs, with minimal back-up and limited or often zero comms, eating military rations and living out of a sleeping bag for days on end.

Will Ferrell and Jake Gyllenhaal might have had to do it like this but if we continued in that manner then *Running Wild* might very quickly run out of guests.

We had to bat smart and adapt the show. To keep it gritty and wild and challenging but make it accessible and fun, and above all short.

Time and experience have taught me one thing when it comes to taking rookies into the wild: it is exhausting for them. The potent, draining mix of adrenalin, fear, awkward terrain and the unknown means that however gym-fit the star is beforehand, they will tire fast. In fact, anything over twenty-four hours is a case of rapidly diminishing returns, drastically reduced concentration and performance, and that's when accidents happen. Not to mention the fact that the whole *Running Wild* experience suddenly becomes less fun. Funny, that.

But that's the nature of the wild: it strips us down fast, lays us bare, and over time it beats the hell out of you. Getting a little bit of that is gold for a TV show. Too much, and the star factor recommendation to other stars (which has become such a huge part of why *Running Wild* has always done so well and has enlisted such big names) disappears. No one wants to recommend doing a show that is going to totally break them and comes with a high chance of getting injured.

If I look back to one of the first episodes of *Running Wild* that we did for NBC, I see all of this in spades. We took the legendary American footballer Deion Sanders, or 'Prime Time' as he is known, on a mission to one of the high mesa wall climbs in Utah. It almost killed him, in more ways than one.

Invariably, the guests on *Running Wild* never really sleep much the night before. They are embarking on something totally new and scary, and they are on their own in a production base in some hotel in the middle of nowhere. All they have seen is a bunch of adventurous-looking dudes running around the lobby in muddy gear and black clothes, dragging ropes and metal climbing tools between rooms and vehicles. None of them speak to the guest. Our team know the protocol. No info, no interaction. Keep the suspense, keep the unknown. The time for 'buddy-buddy' will come once the adventure is over.

I tend to meet the star in their room the evening before, and quickly go through their gear and brief them on what's going on. I don't tell the star much about the plan. Maybe the type of terrain we have ahead, and that I have scouted a rough route from the air by helicopter. I tell them they will be fine, it's all manageable and not to worry. Try and get some sleep. They never do.

We always used to start the next day at dawn. Nowadays we are more likely to start closer to noon. We keep the journeys shorter and sharper, and that way we know that the guest's energy will remain high, and the adrenalin of it all will carry them through the twenty-four hours.

But in season one, Deion was up at dawn.

For our team, fresh from *Man vs. Wild*, it was another day of canyons, steep ravines, hidden drops and improvising our way out of the unknown. All the stuff we love. The only difference being that I now had a rookie guest/superstar with me – but we'd done that before too. Simple stuff.

When we came across a rattlesnake just by this narrow slot canyon, Deion took off running like only Prime Time can. Gone. We eventually got him back, but there was no hiding his raw terror. A lot of shouting followed and it took an age for us to catch the thing. It was all pretty funny.

After an evening of snake for dinner and a cold desert night with not a minute's sleep for the big guy, by dawn he was shattered, cold and wanting out. We've all been there.

I wasn't looking forward to telling him that between us and our extraction point was a 500ft sheer rock wall. So I didn't.

With experience, I have learnt the ability to adapt and tweak these journeys as we go along. If something is truly terrifying or impossible to do, I will adapt the route and keep us moving. It's a balance between pushing people and breaking people. The latter isn't a good thing for anyone.

But on that rock face, the great Prime Time came as close as anyone I have ever seen to total meltdown. The way I have heard him since recount our adventure was priceless.

Not only was he clinging to the rock shaking, but he was refusing to move another inch upwards. The problem was that at this stage, going down was going to be even harder. And slowly Deion realized this.

'Goddamnit, Gryyylls! I'm going to kill you if we ever get out of this mess!' he screamed at me. 'And I truly doubt we ever will at this rate.' He paused. 'How did I ever let you talk me into this?'

Here was one of the greatest American footballers of all time, clinging to this sheer rock face, shaking like a leaf. (He was soon also praying in tongues, which was a moment our crew won't forget in a hurry.) But credit to Prime Time, he never gave up (despite having little choice), and he finally made it to the top through sheer grit and determination.

But for me, it was a day one, lesson one reminder of the old military principle KISS – keep it simple, stupid.

We had planned and executed another journey that was too hard, too long and too terrifying all round for our guest. Much more of this and we'd never even get one season of this show cast and shot.

And as for Prime Time – I love that man. Another giant shoulder that I stand upon. A true legend. But unlikely to be going near any sheer rock faces any time soon.

32

FINDING OUR GROOVE

WE HAVE BEEN so lucky on *Running Wild* to take some truly amazing people on adventures over the years. From Channing Tatum (who we eventually took on a second trip during season five of the series) to Jennifer Lawrence, Zac Efron, Julia Roberts, Ben Stiller, Roger Federer, President Barack Obama and so many others. And season by season, star by star, the show slowly grew in ratings and popularity.

Running Wild came from humble, experimental beginnings and had many a stumble, but soon we started to find our groove. And our former guests and all these stars always loved it. I made sure above everything that we had fun. Yes, there would be some gross and scary moments, and yes fatigue and fear would be part of it, but above all I wanted the experience of the wild to be fun for all of us.

In time, our greatest weapon became the 'star recommendation' by our guests to their Hollywood buddies and in turn to their social media followings, professing that the experience was invariably awesome and empowering. *You gotta do it!*

It was the best marketing the show could ever have had.

The key to its success was that we got to see stars as never before. No make-up, no back-up. Sat round a fire talking about stuff that they really cared about, rather than just promoting their latest movie. The wild does that. It is always revealing, and the journeys themselves created enough trust and fear and fatigue and inspiration to allow our guests to sit on a rainforest floor and talk in a way that few superstars had ever spoken before. It's hard not to be yourself in the outdoors. No disguise. It's called candour. And the wild always inspires it.

The shoots and the seasons of shows became a whirlwind of Central American jungles, hot deserts and high mountains. There was always a lot of travel but the constant for me was our incredible crew. Our friendships were the anchor that kept us sane and carried us through.

Each episode there was a new star – always nervous, always excited. And every time, the shoot seemed to generate this incredible natural energy that was tangible. The recce days were always pretty fun and relaxed, and I got to spend a few hours with the team, either on the ground or in the air, scouting the routes and talking about the guest and the journey. These were often the crew's favourite moments. Pressure off, and lots of laughter. Then suddenly, that evening, the night before the actual journey begins, the atmosphere changes and the star arrives. Game heads on, and it all starts to get more serious. I love that too.

It was always amazing how wide-eyed these celebrities would be about what lay ahead, and one of the big things for us all as a crew was never to get complacent about the locations, the terrain, the dangers – not to mention the very real fatigue and fear that our guests would experience.

As a crew we became so hardened to the job, the travel, the dynamic nature of moving fast over broken ground and steep, difficult terrain that the words from the guests 'You all move so damned fast' became a feature of almost every journey.

It was another reminder for me: keep it steady, be sensitive to how the guest is feeling, judge when to push and when to ease up. It was always a dance but I became pretty good at it.

Amid the busyness of it all I worked hard to carve out and protect as much family time as I could, and that meant school holidays and half-terms with Shara and the boys. The privilege of now producing and making our own show was that I wasn't controlled by anyone else telling me where and when to go and what to do. Shara and I could sit down and plan out the perfect dates and locations to fit in with family life through the year.

It meant always having to be clear on our priorities together: *family first*. There will always be another offer, another request or interview, or another chance to film more. The lesson was to resist it. Protect the most valuable asset you have – and for us that has always been each other.

It meant on occasion turning down some huge names as the dates would clash with half-terms or key school moments like sports days or nativity plays, but with the odd exception it made for a much better and richer life.

It also meant that when I was away filming we would work hard and pack the shows into tight time frames, doubling up on locations and shooting back to back. We became efficient and fast as a team, and we thrived on that energy. At times, this led to a definite increase in pressure and stress on some of the logistics and safety team, but they also became masters of managing that pressure and delivering against the tough backdrop of keeping NBC, the lawyers and our team happy. That wasn't always easy.

As a family, the one thing we would protect above everything was our summer month on the small island in North Wales that Shara and I had bought in our first year of marriage. It became – and still is – our one uninterrupted period where we are all together with few guests and few distractions, beautifully separated from the mainland by 2 miles of wild, tidal sea. We all love it.

I had spotted the island for sale in the back of a magazine that was showcasing private islands around the world. We were twenty-six years old, in year one of marriage, and we had just moved on to our houseboat barge on the Thames in London. We were living week to week, on either the few hundred pounds Shara earned from her teaching assistant work or the proceeds of a few lectures by me to half-filled town halls. The concept of buying an island was crazy and fanciful, especially as all the islands in this magazine were on sale for millions of pounds. Except one.

This one private island in North Wales stood out like a gem. But at a price of £110K we suspected that there must be some catch.

33

ISLAND ASPIRATIONS

THE IDEA OF buying an island only came about because we had saved, borrowed and cobbled together just enough money to buy a small one-bedroom flat in London. The place we'd been looking at was going to cost us £200K. Luckily, in the end we had gone for the more fun and much braver option, and bought our houseboat. It was also the cheaper option, at £100K. In my mind, I figured that meant we had some cash to spare.

So a few months after moving on to our houseboat, when I spotted this private island for sale, I suggested we go on a little road trip to have a look at it. Shara needed quite a lot of convincing. But so had she when we first went to look round the houseboat just before our wedding. It was damp and used mainly as an office by the polar explorer Robert Swan, and I think Shara felt it would be a money pit and miserably cold in winter. I argued it would also be a great adventure.

A month or so after we moved in, Shara was totally hooked, loving every moment of living on the river and being part of this small houseboat community. But trying to suggest that we now also buy an island?

Either way, we drove to North Wales in mid-March in the pouring rain, through the Snowdonia National Park and on towards the coast. The estate agents had told us to take a torch and a crowbar to gain access to the boarded-up old lighthouse keeper's cottage on the top of the island, and gave us the number of Owi, who fished the local waters and kept an eye on the lighthouse from the mainland.

We arrived at his fishing yard and Owi loaded us into the bucket of his digger to try to keep us dry as he drove out into the shallows, and we clambered on to his fishing boat. It wasn't exactly the typical way of being shown around a potential property for sale, but it was fun.

It was still raining as we set off into the grey winter's day and headed out towards the small 20-acre island some 2 miles out to sea.

Owi was (in fact he still is) a man of few words, but he has since become one of our closest friends. Over the years we have seen each other's families grow up, and many of our sons' best memories come from times spent all together. But for now, Owi must have thought, *Who are these naive English folk sat beside me, who would contemplate buying a rock out in the sea that has no power or mains water, that is infested with rats and is regularly smashed by winter storms?* I think he thought we wouldn't last a day.

Owi deposited us on the island and said he would collect us on his return, after he had done his lobster pot round – in six hours. Shara looked at me a bit anxiously, as I grabbed the crowbar and torch and then helped her ashore.

The house, we found out, was totally caved in, with hardly any roof, earth floors, and full of rubble and junk. With no power or water, I knew it would be a near impossible task to renovate, and the amount of materials and labour would be crazy. But the rain cleared and we both sat on the cliff tops and asked the great question: why not?

I'm sure Owi thought that would be the last he would see of us when we finally got back to the mainland, but I felt a real sense of 'meant to be' about the place. It was on the market for £110K and I knew if I could barter it down to £95K then we could potentially afford it. Plus, we would know that for less than the cost of that one-bedroom London flat, we had just bought a three-bedroom houseboat and an amazing (albeit slightly run-down) four-bedroom house on its own private island. What was not to like about that deal?

We then found out that the island was on a very short leasehold, which was a real downer, but again the universe was smiling down on us and a friend recommended someone who knew a freehold lawyer, who in turn told us that purely by chance the law had just changed in Wales and we could actually now legally buy the freehold for 10 per cent of the purchase price of the property. That turned the nature of the investment, and meant we could potentially own the whole island, freehold, and the house, for a tiny fraction of its real value.

We battled hard, negotiated the previous owner down to £95K, lined up the purchase of the freehold, and before we knew it we had bought an island . . .

As for the rats, well that was a battle that only I was ever going to win. I bought an old air rifle and set to work. The island is now 100 per cent rat free.*

---

* The truth is that a local professor from Bangor University, Dick Lock, was the real hero of that escapade. He heard about our problem and kindly offered to bring a bunch of his students over to poison the rats all in one go. He said that there were thousands of them, and that if we left even one pair alive they would return to their previous numbers in a matter of just a few years. He did such a great job, though, and we have never had rats again. Incidentally, they believe the rats first came over in the boxes of supplies that were always going back and forth to supply the lighthouse keeper and his family in the days when the Trinity House lighthouse was manned. Me shooting a few rats with an air rifle really counted for nothing, but it made for a good story to tell our boys growing up, when they asked about the air rifle hanging on the wall with the words 'St Tudwal's Rat Eliminator' etched down the side.

34

RUNNING WILD AT HOME

INITIALLY, SHARA AND I had zero funds to be able to do anything with the island, but that didn't stop us going.

On our first trips, we would go across to the island on the old jet ski that I had previously used to circumnavigate Britain during the first summer of our marriage. Once on the island, we would climb the hill to the cottage and put up our tents outside (and sometimes even inside if the weather was foul). Then we would listen to the huge rats scurrying around all night. No power, no water, no roof. But potential. And masses of it.

Slowly, over time, we saved up enough to be able to pay adventurous friends and builders to go up there and start work. First up was getting a quad bike on to the island so that we could haul supplies up the 350ft of elevation from the jetty to the house, which was obviously sat on the high point of the island next to the lighthouse. Then a generator to power tools. Then we got the first ever Welsh grant for a small wind turbine. Then some small solar panels. Then the lintels, windows, doors and roof. On the list went, but I never grudged putting our earnings into the place. I knew it was going to be an amazing retreat for us as a family over time.

It took us about seven years I reckon, but eventually I was earning enough to be able to make the entire island truly cool. It is now the greatest place that I know of on Earth. Loved by our family, by many

close friends, and by Scouts (I always try to keep a Scout flag fluttering when we're over there, to the family's amusement). But no doubt,
the island is the place where we as a family have spent some of our
happiest times.

Off grid, off comms (for many years at least), few visitors and a lot
of physical living and cold-water swimming. The addition of a pull-
up bar over one of the caves and an old steel slide down one of the
rock faces made the place complete. (It's worth a google of 'Bear
Grylls island slide' if you want a chuckle.) But I'd happily spend all
my days on the island.

I also like the physicality of the place. Everything is hard work.
Even getting a few supplies involves going down the hill, swimming
or rowing out to the RIB, then the rough, wet, 2-mile journey across
to the mainland, anchor down, row the dinghy ashore, hike to the
shops, carry the supplies back to the shore, row out, back into the
RIB, back over, and reverse the process . . . but now uphill.

The addition to the island of an amphibious RIB that has wheels
has made this journey so much easier, but bearing in mind that I am
often going back and forth many times a day, you can see how you get
strong over there. And so do the boys. And that's before we start the
fun all together each day, whether it's climbing, caving, coasteering,
swimming or kayaking. Not to mention paramotoring: I have had
some epic powered paraglider flights taking off from the cliffs on the
island. These have included a few very close calls, such as engines
failing over the sea. Once, Jesse, aged eight, sprinted down the hill to
try and get into the rowing boat to save me from drowning, only for
my backpack engine to restart seconds from watery impact. Or the
time the exhaust broke from its mount and smashed into the props as
I skimmed over the lighthouse. It was the luckiest and best-timed
emergency landing I've ever had as I skidded in just feet from the
edge of a cliff.

The other huge appeal for us as a family was that Shara and I always felt the place was a healthy reminder of real life for the boys. Even though it was surreal life in the sense we were on an island, there was something about just being us with no one around that kept our boys grounded and kept them together. They share a bedroom and clean the loo every day, do the washing-up, fold the clothes, make the bed. All normal, good stuff, but stuff that sometimes, with the privilege of success, can be easy to stop doing. We realized fast as parents bringing up three boys that those are actually the important things. And success must never stop you understanding that truth.

Initially there was some resistance from the boys when we got up to the island, but within a few days we were all back into a good routine of Marmaduke sorting rubbish, Jesse making the beds and hoovering, and Huckleberry cleaning the loos and bathroom. The chores are the same to this day. I love that. And it is amazing how it brings our family together in a totally contrasting way to how wealth, staff or phones so often divide.

Anyway, the point is, the island is special to us. Really special. On so many levels. And those summer holidays from school became ever more sacred to us as a family. And worth protecting.

So, when the occasional *Running Wild* panic came along, with production needing a couple more episodes to fill in where we had maybe cancelled shoots earlier in the year, we soon found a healthy compromise. I'd do the filming, but only if we were within physical sight of the island. And multiple times we pulled that off . . .

35

BACK ON HOME TURF

WE HAVE DONE a bunch of episodes now in North Wales, not only for *Running Wild* but since then for *You vs. Wild* on Netflix and the *Bear Grylls Survival School* series we did for ITV, as well as multiple commercials.

If ever I have used my influence to determine an outcome, it is to decide where and when we film our projects – and in line with *family first*, so often we have brought the world to Wales. I love that. From shooting commercials for Australia's Nature's Own vitamins to Asia's rugged Kyocera mobile phones, every time both commercial partners and *Running Wild* guests love the place. It has it all in terms of wilderness – and of course it's close to home.

I will never forget the *Running Wild* episode with Mel B from the Spice Girls sat on the top of a 300ft sea cliff, watching the sun setting, with the lighthouse on our little island beginning to flash away in the far distance. For me, it was a spectacular way to see my worlds of work and family coming together.

We were now filming a show we owned, and doing it for NBC, the biggest network in America. We were then selling the episodes for secondary viewing to Discovery Channel, getting paid ten times my *Man vs. Wild* fees, and being treated so well by both those partners. We were taking superstar guests on adventures, in a safe and manageable way, for short time frames – and to top it all, I could see home on the horizon. Good job, team. Yes, we had to fight a few battles, but we got there.

Mel B was always going to be a fun guest, and it is always a nice change to film with fellow Brits. However much fun I have with many of these other big Hollywood actors – and don't get me wrong, we have had so many laugh-out-loud moments, on and off camera, when filming with US stars – there is always something special about filming with British guests that is hard to quantify. It's the banter and the subtle mick-taking that is so good to be around, and which has always been such a big part of our own crew's existence. It's a harder form of humour to carry with US stars. That sort of atmosphere on a shoot always feels like home.

Mel B brought it in spades. She was brilliant. I will never forget her urinating on my hand after I got stung by a jellyfish. Classic moments. You've got to laugh at life sometimes, eh?

Mel was such a good guest for us, for so many reasons. The Spice Girls had been the biggest girl band of all time, all over the world, and that label will follow them for ever. And that is the magic of *Running Wild*: every guest brings their own unique audience and fans to the show, which means it doesn't always hang on my shoulders, or how extreme or crazy the adventures are. The star power of each guest has always been a big part of why ratings have been so high, and I'm forever grateful for that.

Mel B was a textbook shoot for us – always game for an adventure and brutally honest about her life and struggles. Those are the other key elements of the show's success – willingness and authenticity. And then what's not to love about ending any adventure with an under-slung jet ski being dropped off by helicopter for us as we emerged from a sea cave to swim out to it and ride off . . . great moments. And, of course, I was then home on the island by lunchtime.

Another time, we brought the Oscar-winning actor Kate Winslet to North Wales – but this episode we shot in the mountains. Still, we managed to finish the shoot on an amazing peninsula of wild terrain that I had actually purchased with some friends a few years earlier, as a way of protecting the headland from ever being developed and built on.

For us to rappel down the cliffs of that headland and swim out to our amphibious RIB for the extraction was special. Kate loved it all and has remained a good friend. Another fellow Brit who would take the mick out of me as much as I did her. I said our adventure was like camping with Mary Poppins, and that was pretty accurate. She was relentlessly positive, brave and good-humoured – all such strong survivor traits – and I would have her on my team if we were stranded in some jungle any day.

In fact, it's strange how the women we take on *Running Wild* so often outshine the men. I don't know what it is. Not all the time, but often. I guess we are all guilty of judging people on appearances at times, and as a crew of predominantly big, butch, male mountain guides (with the exception of some awesome women who, once again, often outshine all of us in terms of attitude and workload), we can be an intimidating bunch to encounter on a trail.

But time and time again we see such brilliantly strong, positive performances on our trips from the least imposing of actors. And I love that. It's a powerful reminder, not only to us but also to the 1.2 billion around the world who watch the shows, that strength and character and survival spirit have nothing at all to do with gender, age, appearance or whatever society-driven stereotype we might be tempted to pre-judge people on.

In fact, strength, character and survival spirit are not exclusive to any group of people. They are in us all. It's a choice to find those qualities, to test them, and to let them shine. That's what I love most about survival. It is already in us.

36

RUNNING WILD, RUNNING FAST

THERE IS NO doubt that taking both President Obama and Prime Minister Narendra Modi of India away on adventures have been high points in my work life, and I don't want to take anything away from that privilege. But for me, personally, in terms of fun, laughter and loving the moment, those shows were a nightmare. In truth, I was simply too darned scared to enjoy either of them.

The fact that both Obama and Modi were sitting global leaders of two of the most powerful nations on Earth meant that the security, logistics and protocols on those journeys were highly restrictive. We also had the added pressure of working to incredibly tight schedules, which meant we had to make the filming even faster and more efficient. Not easy. Everyone was constantly on edge.

It made it hard to relax and enjoy the experience. It's like playing the final quarter of the Super Bowl or the fifth set in a Wimbledon final. You've got one chance. Don't screw it up.

In the weeks before filming, the networks were also endlessly sending me suggestions of chat points, most of which were way off mark. I knew that the conversations would need to be pitched just right, respectful but light, inquisitive yet not intended to make the President or Prime Minister defensive.

Then, of course, there was the whole dynamic of having a

real-time Hindi translation buzzing away in my earpiece, as was the case with the Indian Prime Minister.

None of the above was straightforward, and it didn't leave a lot of room for levity. It was more a sense of 'deliver, or it's over'.

Yet the beating heart of *Running Wild*, why it works, is that the show is always rooted in fun – whether with the crew or the guests. Laughter is such a key part of life, seeking out those moments when you just can't keep it inside any longer. But all too often filming can stifle that spirit, partly because everyone is that little bit more self-conscious. Myself included.

It's hard to totally forget you are on camera, and I'm always aware we are there to do a job. *Get on, climb that peak, and chat coherently – and together let's make sure everyone is safe, getting their shots as planned, and staying on time and to schedule.* It makes moments of total abandon or hilarity harder to come by. But when they do come, they are good.

I remember the show we did in Sardinia, for example, not so much because of the adventure but because of a goat.

By the time we had all scuba-dived inside a wave-beaten sea cave and then made it to the top of a 500ft cliff, we were all ready to have a drink and a rest. I suddenly spotted a mountain goat grazing among the bushes and after a bit of a tussle I managed to catch it. Jokingly, I then suggested sucking the goat's nipples to get some milk.

Now obviously, we could have just squeezed the nipples and filled our canteens, but somehow I ended up missing out the middleman and sucking straight from the teat. The goat didn't seem to mind it too much, although she did try to kick me square in the chops when I bit down a little too hard.

I remember seeing Mungo's shoulders shaking with laughter as he tried to keep the camera steady, and soon we all had tears coming down our cheeks, as the nipple started spraying milk everywhere. Eventually, we just had to stop filming.

I love times like this. Unplanned childish humour, messing around with good buddies. It keeps life simple in a world where things can get heavy all too quickly. Because, of course, when it comes to animals and food sources, everyone has an opinion.

This sort of interaction is nowadays about as far as we go with animals. Popular culture has changed so much since I started filming *Man vs. Wild*. In the early days I was always hunting down some croc or pig or snake. I was after meat – and inevitably blood and guts became a visceral part of showing what real survival involves. It wasn't always pretty, but it was what many of the viewers liked to learn about. Today, though, there is an increased sensitivity to animals. And that's not a bad thing. I look back and I don't think I always got it right. But those of us on the early shows never really thought that much about it. We were simply showing what you would need to do if you were stranded alone and had to hunt to survive.

But over time I became aware we needed to change and adapt. It is why now, when we need food on any of the shows, we make do with old carcasses, a few dead pigeons, digging for worms, or munching on a handful of grubs and critters. People get the idea. And no animal gets harmed in the process. That's a good thing.

As Jesse said to me recently: 'You're the poacher turned gamekeeper. You used to be a killer, now you're a conservationist.'

Maybe a little harsh, but I'll take it as positive progress.

37

IMPROVISE, ADAPT, OVERCOME

PEOPLE OFTEN ASK if all these *Running Wild* stars become great friends afterwards, but the truth is that people tend to come along on the show and have a hugely empowering and positive time, rave about it for a few weeks to family and friends, and to me, but then move on to the next project. I think Hollywood is notorious for this: you become very close during filming but then you move on. I don't like that dynamic very much.

I remember at first I found it quite weird. I come from a world and a background where as a team we stay together through everything. We build friendships for life and we always keep in touch. The friends I have like this aren't many but they are good and true. I like that.

When *Running Wild* started and we would go through so many great moments and adventures with these stars, I found it strange how fast they would move on in terms of their relationships. Not just with me but with our crew with whom they had also bonded. It felt a bit fickle.

That's not to say I don't keep a positive link with almost all of them, and I will often check in and get texts if something great has gone down in either of our lives, but it is just that they don't all become the friends that maybe you might expect. But that's Hollywood, as they say.

In fact, out of *Running Wild*'s many guests, the ones I keep in touch with amount to a small number. But I love those guys. Channing

Tatum would be one, as would the likes of Ben Stiller, Roger Federer, Kate Hudson and Kate Winslet.

In life you generally only need a handful of good friends, and they certainly don't need to be big superstars to be loyal, good buddies. Thank goodness.

I remember Armie Hammer telling me, some time after our adventure in Sardinia, that the day and journey we went through together had been so mind-blowing in intensity, that it was hard for him to describe it to his family when he got home. It's always good to remind ourselves as a crew that what we all consider routine really isn't routine at all.

Armie had arrived with a high level of scuba experience, having done a bunch of deep and committing dives in various locations around the world. He wasn't just a regular scuba dude, he was well trained and experienced. It was why I felt comfortable pushing the boundaries a little, in terms of our planned infiltration along the rugged coastline of Sardinia.

The intention was to drop from the side of a RIB at a point we had scouted at the base of some committing and high sea cliffs. We had been told by local divers about a sea cave at the base of the cliff, with a small venting blow hole that could be accessed from deep within the rock face. That blow hole would lead up to a small ledge some 50ft up the cliff, from where we could begin the climb.

All possible – but I knew it would require some technical scuba skill. Armie, the crew and I would be operating in narrow, dark tunnels, sub-surface, then would have to sink and cache gear before attaching a rope ladder to the narrow blow hole above, and climbing it – all in strong tidal flow.

That's why we had warned Armie that this was going to be an adventure.

As it happened, the conditions on the day were rough, with strong mistral winds blowing, making the waves and white water at the base of the cliffs horrendous. I knew that would mean huge hydraulic suction of the water within the subterranean tunnels that we were planning on operating inside. The local dive crew deemed it too dangerous, but with a slimmed-down crew we decided as a team we would at least have a look at it. We could always bail and go to plan B and a secondary route on to the cliff faces if needed.

We got geared up and then dived down into the swell, finning towards this narrow black hole at the base of the cliffs. White-water waves were barrelling just metres above us. Two of our crew had regulator issues with their gear, exacerbated by the sea state, and had to retreat to the RIB – but Armie, Dan our camera operator and I just managed to get into the tunnel by the skin of our teeth, with me leading the way with a torch.

It was instantly like being submerged in a pitch-black washing machine, and in such situations keeping calm, aware and alert is everything. One small mistake would have had big consequences. The three of us just managed to get through the narrow squeeze into a small chamber, where we could brace ourselves together and begin to work our plan. This involved the not uncomplicated procedure of hauling gear through the tunnel, and attaching lines, poles and hooks in place in order to establish a vertical free ascent to the blow hole above. All done in the dark by torchlight.

This then had to be followed by cacheing the scuba gear on the sea floor and coming back up with only inches of head space above in which to breathe.

You get the idea. It is many people's, including Shara's, worst idea of a nightmare situation. Dark, confined, underwater spaces, with limited air and huge wave pressure sucking you back and forth, dragging you against sharp rocks on the tunnel roof above.

Anyway, we did it, and all went to plan. Armie did everything perfectly and was calmness personified in quite a high-octane, committing situation. But he said to me many months afterwards that even with all his dive experience and training he had never done anything so truly terrifying and totally dangerous in all his life. In his words: 'the risk factor was off the scale'.

But I look back and don't think it was so bad. We knew the route, we knew the drills, we could manage the risks and had plans for gear failure, human error and nature's unknowns. But to be able to operate in this fashion you have to be both current in your skills and confident in your emergency procedures. The mantra we go by is: *Anticipate the worst, and plan for the unexpected.*

As a team, by the time Armie and I had got through that tunnel we didn't give the danger behind us much of a passing thought. And that is so common for us in our *Running Wild* life. There are moments of great intensity and risk that require optimum focus, but as a team we tend simply to do our jobs and deliver, then move on. Within minutes we can go from high alert to total laughter. And by the end of the day you've forgotten it all. You're already thinking about tomorrow.

And that's how it was with Armie. We got through the sea cave, we cracked on with the climb, and a few other bits and bobs, and finished with a huge rappel down another cliff face on to a jet ski for our extraction.

Still soaking wet but now off camera, we jumped into a helicopter inbound to the island's commercial airport. There we were met by some officials, raced through customs and security via a few back channels, and eventually just managed to get on our British Airways flight with seconds to spare before the plane door closed.

Armie collapsed in his seat next to me, and said, 'Jesus wept. Bear, tell me not every day in your life is like that . . . I'm exhausted!'

But the truth is that when we are filming, days are often like that. As a crew, we definitely become a little hardened to the adventure, the adrenalin and that relentless need to adapt and improvise.

But for each of us, that's what we pride ourselves on being able to do, against the odds, over and over again.

Improvise. Adapt. Overcome.

38

RUNNING WILD FAVOURITES

WHICH FINALLY LEADS me on to my top two episodes of *Running Wild*. First up would be the Red Nose Day Special we did for NBC.

Richard Curtis, the legendary founder of Comic Relief and the writer of so many iconic movies, approached us to ask if we would consider doing a *Running Wild Special* as the lead show for their on-air fundraising evening across America. He then asked if we would consider taking the Oscar-winning actor Julia Roberts somewhere in Africa on an adventure with a mission. It wasn't a hard decision.

Together we planned a journey into the Kenyan bush where Julia and I would deliver vaccines to a remote community. Beyond the actual mission itself, the goal was simple: for NBC to use this one-off episode to help raise tens of millions for dying children around the world.

As a team, we were in.

Julia is one of those iconic Hollywood superstars that I had grown up watching. *Pretty Woman*, *Notting Hill*, the *Ocean's* films, *Erin Brockovich* – I was a fan from way back. It made me a little nervous. I always feel like this, even to this day, around big stars. I hope I don't ever grow out of that.

The scouting for this journey hadn't been straightforward, and the crew at one point had to deal with coming under small-arms fire between two warring tribes in a remote part of the country. That always spices up anyone's day, to find yourself in a desert gully with rounds and ricochets going off all around, and realizing that you're right in the middle of the gunfight.

Credit to the team that they kept cool heads – they sat tight and stayed low, and let the battle pass on by – but it is another example of what our team, on occasion, have encountered when simply scouting terrain and planning a journey.

We had firmly decided not to mention firefights to any of Julia Roberts' team, and by the time Julia was in Kenya we were prepped and good to go.

What's always so fun for me on *Running Wild* is taking these huge global stars back to basics within about thirty seconds of meeting them. It is always priceless to see, and invariably a good way to bond at the start of an adventure.

We knew we would have a fixed-wing insertion to where we were going to be filming, and when a classic African Twin Otter bush plane arrived on the dusty desert runway, we got to work making it fit for purpose. To our team, that meant ripping all the passenger seats out, filling the back with crates of chickens and bales of hay, and basically making it look as rickety and rustic as possible. And in this regard we are always playing to our crew's strengths. By the time we were due to take off and head out to collect Julia, the plane looked like it had been mothballed in a farmyard for the last decade.

A good start is always important, and in our *Running Wild* world, the rougher the whole plan and gear look, the better. Even if, as a crew, we know it's all good and fit for purpose.

The beauty of operating like this is that it instantly removes all the formal airs and graces these big stars might arrive with, and in a heartbeat it turns the entire experience into an adventure. If we get it right, the tone is then set for the whole of the upcoming journey.

We landed on this tiny, remote bush airstrip, we kept the props turning and burning at half power, and I leapt out to meet Julia. We could hardly even talk above the noise of the prop wash, so I just grabbed her by the hand and raced her to the rear cargo door.

The dust was blasting everywhere, the plane was shaking, the chickens were going crazy inside and hay was blowing all over the place. Awesome. Julia had one hand on her hat and was screaming with that dizzy mix of horror, delight and fear. I bundled her through the hatch, gave the pilot the thumbs up, and as the plane hurtled off down the runway the G-force threw us backwards into a heap of hay, chickens and boxes.

The journey had begun and we were all guns go.

I love this way of working, and from then on we had such a fun time.

From the plane, we transferred to an old Land Rover, to get us a little closer to our final destination – the Kenyan village where we would be delivering the vaccines. En route we got a flat tyre that Julia helped repair, with her rolling around in the dirt and dung of the African bush, and we talked about life, family and all that good stuff. Then, when the track ended, we carried on by foot, trekking through the 40°C African heat. Despite the grime, and fatigue, Julia was awesome, and was always laughing. It was *Running Wild* as it always should be: the team were in their zone and the star was perfectly terrified, exhausted and excited all at the same time.

An old hemp bridge crossing over a crocodile-filled river took Julia to her edge, but again, every good episode should have those moments. The dance is knowing how to get the stars into it without too much thought, and then to help guide them through it with kindness and encouragement. Invariably, the guests are so proud of themselves when they conquer their fears, and rightly so. Massive cliff faces and deadly animals are always a little scary. It is why I usually feel the wild does most of the job for me.

Julia did so well, and by the time we had made camp, built a fire and eaten the remnants of some old goat head that we sourced en route from a local Masai Mara tribeswoman, Julia was dead tired. The guests always are.

The exception being the occasional character like Scott Eastwood or Cara Delevingne, who were only too happy to break into our crew's 'emergency' flask of rum and sit and watch the moon and stars dance across the horizon. The blissful lack of light pollution at our locations makes those evenings camping out so special. Nature's TV at its best.

39

MEETING YOUR HEROES

THE GREAT STRENGTH of *Running Wild* is that it attracts so many different audience types, and not just the adventure lovers. Unlike so many outdoor shows, we get everyone from the teenage fans of a pop-star guest to sports fans and movie fans. The great value of this is that every episode has its own audience, often tuning in not for me or the adventure but to see their hero out in the wild, in such a revealing, honest way. That was always gold for the series' ratings and kept the show continually refreshed with new fans.

Of course, each of us on the crew also had different guests that we would get particularly excited by. Sometimes I have taken younger stars away without being aware of quite how huge a following they might have. Take the boy band superstar Nick Jonas, or Oscar-winning actor Brie Larson. Both huge stars in their genres, and in the case of Brie we shot her jungle episode before her blockbuster Captain Marvel movies had even aired. Such things have meant I don't always appreciate the popularity of the person I am with. I think that can be a good thing, though, as it means I don't approach them with too much fan baggage. Everyone is just a regular person on *Running Wild*. I like that.

When we started to do the show in China, with Chinese celebrities, I might never have even heard of the star we were meeting beforehand. Yet they would have hundreds of millions of social media followers and untold numbers of young teenage fans around the world.

To me, it was always a strong reminder of what a fickle world the celebrity cult can be. Does it really matter if I've heard of someone before I meet them? The true test of a human being is how they hold themselves and treat others, particularly if they possess fame and success. The great currency of humankind is: are you humble, are you kind, and have you come through a battle to be here? Those are the things I am impressed by in people.

Other times – as was the case with Julia Roberts – we would be taking someone away whom I had grown up watching in films, and it was fun to see the crew, all much the same age, get so excited beforehand. Myself included. (Although often our boys back home wouldn't know them. Such is the fleeting nature of fame, I guess.) It was funny how much more nervous I always was in those instances. It's hard not to get dazzled by our heroes sometimes.

The one guest I had wanted to secure from the very start of *Running Wild* was the legendary Roger Federer, arguably the greatest tennis player ever to have lived. He had said no to us a few times over the seasons, always promising to do it when he retired. Then, when he returned from his astonishing eighteenth major title win at the Australian Open, he was fired up and on a natural high. When we connected by text and realized we were both in Switzerland at the same time, I gave him a solid push to take the chance now and do an adventure together in the Swiss mountains. We could, I reasoned with him, do it super close to where he lived.

Good work, BG. Make it super easy for him to say yes.

And I have got to admit, when he agreed, I did have a little moment.

From the get-go, and all through the planning of the Federer journey, the crew took the mick out of me ruthlessly.

'Bear, we thought that you and Roger could hold hands for a

slow-motion opening of the show, walking through the Swiss Alps together . . . what do you think?'

I had a barrage of texts and emails like this. And even on the shoot day itself, as I stood waiting to greet him on some remote mountain road, feeling decidedly giddy, the jokes kept coming.

'Bear,' Mungo would mumble from behind his camera, 'you're looking a little flushed, brother. You OK?'

They always say don't meet your heroes, but in this instance they were wrong, and both Roger and our journey were awesome. He was the consummate gentleman throughout, despite me tearing his new Mercedes apart to gather supplies and fuel that we might need on the trip, and he threw himself into everything. Sometimes literally.

The first thing we did was drop off this sheer mountain road, with the rope tied off to the wheel of his car, descending down a frozen, vertical waterfall on crampons. Not easy for a rookie new to ice climbing.

Normally wearing these razor-sharp metal points fixed to mountain boots and navigating sheer sheet ice can be a recipe for some slips and gashed gaiters. But Roger had the balance of a bird, and picked up the technique so fast. Before we knew it we were down into this deep snow chasm and on our way.

Roger was definitely squeamish about pulling out the frigid remains of a fish from the side of a frozen stream, but he relished chatting around a little fire in a tiny cave, talking about life, his career and family. We even got to pee on the fire together, which, all things considered, is the ultimate bonding moment for any bromance.

By this stage of the journey the sun had set, it was getting markedly colder down in this snowy canyon, and Roger's hands were beginning to lose sensation. Not good for a man who needs good sensation in his hands more than most mere mortals.

I knew we needed to keep the pace and keep working hard, and that it wasn't going to get any warmer on its own down there. We improvised a type of lasso with a tennis ball and some paracord, and Roger deftly managed to get it looped up and around the roots of a small tree some halfway up this cliff face. First part done. Now we had to get up this precipice ourselves.

Roger, for all his fearlessness on court, is much more nervous with heights, especially when climbing such awkward ground in the freezing cold. But he muscled it out and soon we were near the top.

I had one final thing I wanted to do, though. And this would be the perfect moment. He was tired, shaking a little with cold and fear, and could hardly feel his fingers.

If ever I was going to beat the GOAT at a game of mini table tennis, it was now.

40

GAME TIME

ROGER LOOKED AT me a little confused.

'What . . . here?' he said. 'In five feet of snow, on the edge of a precipice?'

'Right here, right now,' I replied. 'Mano a mano. Winner takes all.'

The GOAT smiled. It was game time.

Now, we have done a lot of fun stuff over the years on *Running Wild*, from cool techniques to 'tickle' and catch fish to improvising crampons to cross glacial ice faces, even bootlegging one-man self-rescue systems for escaping a crevasse. I love all this stuff. But the ping-pong table folded up in my backpack in order to challenge Roger on this mountain peak was always going to be a special moment.

Talking of cool techniques, I remember once rappelling off a huge cliff secured only to a frozen chocolate bar – which, by the way, worked like a dream in the sub-zero Arctic conditions. I had simply wrapped the rope around the frozen chocolate and buried it in the snow as a strong point. Job done.

On another occasion, I lowered a Hollywood star and myself off a precipice with our rope secured only to a mobile phone wedged in a crack. Although that one was a bit more touch and go to be fair.

We had set the phone to vibrate mode on a seven-minute timer, which I had planned would give us just long enough to reach the bottom of the ravine before the phone would start to shake itself free from the knot system, thus allowing the rope to be retrievable by us from the bottom. That one worked perfectly, but sadly the phone (one of our emergency comms) died when it fell down along with the rope. But the anchor principle was great, albeit definitely a little sketchy and maybe not one to try at home.

Then again, many of our techniques could carry that warning. It is why I am always quick to add that 'this should only be a last resort' when setting such systems in place. My point is always that ingenuity and resourcefulness are at the very heart of all self-rescue survival scenarios. So you can survive with very little expert gear, if you're smart.

I also hasten to add that whenever we do use really precarious anchor points like this, it is more about showing clever, innovative ways of staying alive, and should never be all you are depending upon – unless there is truly no other option. This is why I always back these innovative systems up with proper gear. Interestingly, the stars hardly ever notice me or the team putting an extra rope or anchor in while they are getting set to rappel. I guess they are so in the moment they're not always aware of what else we are doing quickly at that point. But as they say, safety never takes a day off.

It is always amazing to see in action the trust the stars place in us, and I never take that for granted.

I could also include in this list of innovations the time I once made fire from urine with the actor and wrestler Dave Bautista. That was something I had wanted to show on camera for a while, and it worked pretty well. One clear bag of urine, a hot sun, and some good tinder. Then refract the sunlight through the pee bag on to one small hot spot and wait for smoke and then a small flame to appear. Bingo.

Anyway, back to the table tennis.

We assembled the mini table that I had pulled out of my pack and got ready for battle.

'First to eleven. You want to serve?' I asked.

'What, no warm-up?' Roger queried.

'Straight in,' I replied.

Who did this guy think I was? I'd waited my whole life to have a chance to take on Roger in a match and there wasn't a hope in hell of me affording him the luxury of a warm-up.

I also wasn't going to let on to Roger that I'd been practising on this mini table for months. Because by all measures, mini table tennis sets are tricky, awkward things to play on at the best of times. I figured my no-warm-up tactic should at least gain me a couple of early points.

The crew and the two of us settled down for the match of my life.

And as hoped, I soon found myself 7-3 up against the greatest of all time.

All was going sweetly to plan.

Until Roger started to get his eye in.

I hadn't anticipated this happening so quickly, and before I knew it we were 7-7, then 9-9.

I took a deep breath and looked across at Scott, one of our safety team, tucked back behind a tree. He was discreetly holding a rope that was secured to Roger's harness to stop him sliding off the cliff face in case he leant out too far for the ball.

Scott is a great buddy and a veteran of so many of our adventures from the early days of *Man vs. Wild*, having joined our team on the Sumatra episodes of season two. He has been with me pretty much ever since. As a former Physical Training Instructor with the Royal Marines Commandos and then a policeman, Scott has become fundamental to our core safety team. He also knew how much this moment would mean to me.

I looked across at Scott, who grimaced back. I could feel him willing me on. This match was suddenly very tight.

Rog hit a perfect backhand that zipped over the net for an annoying winner. 10-9.

Hold tight, BG . . . you can still do this.

That thought was a little premature, though. I suddenly felt my elbow go to jelly and my vision blur as a serious dose of the yips overcame me. Now I knew what many of the great RF opponents have felt over the years. And it didn't look like I was about to fare much better.

And sure enough, after an epic final rally, I buckled a shot and wobbled it long. Game, set and match, Roger Federer.

After all that preparation, I blew it.

By the time NBC edited the final show, along with Wimbledon green graphics, scoreboard and commentary, it was a classic bit of *Running Wild* fun and adventure. And true to form, we made the guest the hero. At least, that was my reasoning to the crew when mercilessly teased about my defeat later that night over supper.

Yep. Always make the guest the winner.

41

THE GREAT LEVELLER

BOTH THESE STARS, Julia Roberts and Roger Federer, are such big global icons, and it is always strange to meet your heroes. But in many ways, now that I know them and can consider them friends, they are even more impressive to me. That's surely the real mark of a true superstar. To be able to say that once you get past the dazzle, and get to know the real person, behind it all, behind the fame and achievements and stardom, they are just like you and me.

Still to this day I sometimes have to pinch myself and ask how, as a team, we ever got into a position where we were asked to take such incredible legends away on adventures, and to give them experiences that money can't buy. But, of course, the real unsung star in all of this is the wild. And that's how it should always be.

The wild tests us, and asks what we are made of. It demands courage and commitment if we are to survive, and rewards kindness and resilience in the big moments. You only have to look at the times when things go wrong. That's when we find out what people are really made of. How we all react in the pressure moments. As they say: no pressure, no diamonds.

The challenge with taking any sort of rookie on an adventure is that our team has to contingency-plan twice as hard. Because rookie means one thing: increased risk.

And the one factor that is hard to control above any other is the weather.

Ben Stiller was the first Hollywood star we brought to the UK, and we chose the Isle of Skye as a great wilderness island location. The fact that the Cuillin mountains are notorious peaks known for stormy weather was always going to make the shoot a challenging one. The Cuillins are where many of the UK Royal Marines Commandos do their mountain training, and with good reason.

They say if you can survive and climb there, you can survive and climb anywhere. It's that dangerous mix of driving rain and wind straight off the North Atlantic Ocean that makes it so challenging. Staying 'dry in Skye', the Commandos say, is near impossible.

Sure enough, the weather delivered, and I will never forget flying in the chopper to collect Ben from just outside this old whisky distillery. It was a day of classic Scottish drizzle and grey foreboding skies – and the outlook was bleak, with ever-deteriorating conditions at higher elevations.

Ben was dressed in a baseball cap and sunglasses, and looked like he had just popped out from watching an LA Dodgers game – albeit a little more wrapped up. Within minutes of meeting and flying off together on the chopper skids, we were dropped just under the cloud base, barely a quarter of the way up one of the Cuillin peaks, in a torrential downpour.

We huddled against the beating rain and I shouted over the wind that our direction of travel would first take us up and over the summit. Ben looked worryingly up into the clouds, and with a nervous shake of his head and his wry smile we headed up.

One of the things I love about *Running Wild* is how fast the wild and the weather can change someone, both in appearance and in character. And Ben experienced this in spades. By the end of our journey, some thirty-six hours later, having battled horrendous conditions, faced some serious drop-offs into deep ravines, scaled thousands of feet up and down, and slept in a dark, dank cave, he was truly unrecognizable from the Hollywood comedy legend who had stood at that distant distillery waiting for me.

Both of us were shivering as we held the tartan material from two old kilts tight around our necks. We had grabbed the kilts from the back of the seaplane that had been our method of extraction earlier that final day.

But what Ben also wore was a smile that was as wide as his face, eyes shining with pride at what he had endured and achieved. This is why the wild always bonds people so powerfully. It is hard to be anything but yourself when you are up against horrendous conditions on a mountain – and the true side of a person is always so much more attractive than the more polished versions we all at times like to portray.

Then there was the horrific storm we had in the Pyrenees with the Netflix sensation Uzo Aduba. For an LA lady who is most certainly not accustomed to wild weather, it reached a point where it was almost impossible even to hear each other over the torrential conditions – rain that truly rivalled any jungle monsoon I have known. All we could do was hold each other under a tiny poncho and laugh.

Moments like that, with her beaming smile and infectious laugh, are special. And if the wild does anything for us, it unites us. Regardless of nationality, colour or religion, all of us stand in awe at the power and magnitude of Mother Nature. And that always brings people together. The great leveller.

Then again, some people just plain don't like bad weather. And of course, when ill-prepared and alone, storms can be truly terrifying.

I remember taking the basketball legend and Hall of Famer Shaquille O'Neal into the Appalachian mountains in upstate New York, not far from Mount Washington where the highest wind speed in human history was recorded at 231mph.

I also took the actor Don Cheadle, from the Marvel movie franchise, to that area, although Don got luckier with the weather. Instead he got a busted rib when he leapt off an old railcar a little prematurely with me, but that's another story.

Shaq was a gem, and at over 7ft tall and 325lb, he is one big unit. The boots we got him for the adventure were more like boats and the only gloves that we could find to fit him were oven mitts. He saw the funny side and went along with everything we did. Even eating the rotting placenta from a moose carcass with me.

I had also never seen a set of climbing ropes go so bar tight as the time I had him hanging off a cliff face secured to a huge oak tree. I wasn't sure what part of the system was going to break first under the strain: the tree, the rope or Shaq's battle against gravity. It was touch and go on all fronts.

When the weather turned that evening, I figured the fastest and simplest way to keep the giant man dry was to leave him where he was, slumped against a tree, and simply to pile up dead leaves all around and over him. The leaves kept him perfectly waterproofed and insulated, and with only his bald head poking out of the top he was the picture of wilderness innovation and positive morale.

'Just give me my hookah pipe to smoke once you stop filming,' he said, 'and I'll be happy as you like for the night.'

And he was.

Legend.

42

CLOSE SHAVES

THE VERY DYNAMIC nature of *Running Wild* means that inevitably, over time, some stuff goes wrong. Our job is to plan for that and to minimize the consequences. I like to think our team do this better than anyone else in TV.

You might call that a bold claim, but I would challenge you to find better. We know how to operate under pressure, in the toughest of terrains, and how to do it in a low-key, egoless fashion that draws on the wealth of experience I and the safety team have built up from a lifetime of adventures – and misadventures. After all, isn't it when things go wrong that we always learn the most?

I'll always remember the journey we did with the Hollywood actor James Marsden, from the movie *The Notebook* and the hit show *Westworld*. When we got halfway down this steep, scree rock face in Nevada, negotiating an awkward undercut shelf on all fours with a sheer drop-off only inches to our right, James turned to me and just shook his head, chuckling to himself.

'You good, buddy?' I asked, as we both shuffled along.

'Yep, but I just can't believe we are really doing this.'

Still on all fours, he looked down to one side at the cliff face only inches from him.

'What do you mean?' I replied, tightening the short rope I had between us.

'It really is just us,' he said, 'and it really is just f***ing dangerous . . .'

Often, I think that these stars arrive expecting it all to be like a movie set, with stuntmen checking everything, and catering trucks and make-up, and all the fluff that normally surrounds a TV show or movie production.

When they find out it is just them and me, and a small team of mountain guides to help keep the crew safe and to be an extra set of hands if needed, it is a shock. Then, when they realize we really are camping out, and we really are in the middle of nowhere, and that danger is a constant journeyman with us on these adventures, there is often a moment of realization that is priceless to see.

But invariably it is why the stars love it. For just this one moment in time they really are doing it and they really do need to keep hold of that rope for themselves.

The next day, on the final rappel that James and I did down this massive 300ft vertical rock face, I saw a raw fear in his eyes that is impossible to disguise. But he did it. He kept moving, kept trusting both himself and me, and just did it. And that's always the point. Can these people, usually so protected and privileged, find a way over their fears and nerves and just do it? When the answer is yes, it's amazing to see.

I often say to Shara and my young boys that it is our willingness to face the difficult stuff, repeatedly in life, that makes us strong. It's the failures and the times we get knocked down yet have to stagger back to our feet and go again that build our internal muscle, our inner grit. And that muscle is everything in life.

Whether you're an actor in Hollywood or a regular Joe, however successful you are, you have to get used to facing difficult, scary situations in life. Perhaps for some of our guests it's auditions and dealing with rejection, or sporting defeats and a loss of form. Whatever the setback, rebounding is a core skill for success, in all walks of life. This is why so often these stars are good at overcoming fears. They are used to it. Just in a different arena.

Probably the closest call we've had, though, was with Michelle Rodriguez, the legendary actor from the *Fast & Furious* franchise.

We had planned to skydive in together, to the Red Rock desert canyons outside Vegas, and then start the journey from this wide valley of scrubland we had earmarked as a good LZ.

When I say skydive together, what I really mean is that I was to throw her out of the plane and then dive down directly after her. The technique requires some basic parachute training with one of our team beforehand, but as long as the person is on a radio we can generally talk them down safely once they are under canopy.

We've done it now a few times, with Zac Efron, Channing Tatum and the free solo climber Alex Honnold. It didn't go wrong for any of them. Then again, none of those guys have the wild child reputation that Michelle Rodriguez had. If ever there was a free-spirited girl, up for danger and an adventure, it was her.

All went fine with the training, and after picking her up from this desert airfield, at 12,000ft I opened the plane's door, helped her to the edge and nudged her out with a smile and a shove. All I heard was her screaming with delight at the adrenalin injection.

I dived out after her, saw her chute open, then deployed at a much lower altitude so I could be on the ground before her, in order to help radio her in.

The problem was that her primary radio wasn't working, and her secondary back-up was full of static. Despite all my best efforts, she was flying blind.

Pretty quickly we realized she was heading for a 100ft sandy cliff bluff at the edge of the valley.

'Turn now. Come on, Michelle. Turn that canopy to the left,' I was shouting into the handset – to no avail.

The moment we saw her crash into the side of the valley was a terrible one.

We all just started running across the desert to try and reach her. It took us at least ten minutes of sprinting and then scrambling up this steep desert bluff to get within reach. We were all pouring with sweat and expecting the worst as we started to climb up to where she was hanging in her harness off the side of the valley. Somehow a rock had caught her chute, and she was now suspended just above us, swinging under the canopy.

'We've got you,' I shouted. 'It's going to be fine.'

'That was a rush!' she shouted back.

By the time we got her down and checked her out, we could see her trousers and shirt were ripped and trickles of blood running down her leg and forearms.

'We can stop right now, Michelle, if you're hurt,' I told her calmly. 'That was quite a crash.'

'Nah,' she announced. 'It's just a scratch. Let's get going.'

And that was that. With her hair full of desert dust and the blood already crusted dry on her arm under the desert heat, we just cracked on.

Even later on, when we caught a mouse in a shady ravine and stewed it in our combined urine, she hardly batted an eyelid.

She simply looked at me with a grin and a shake of her head. 'If needs must, I'm in,' she said before crunching through the bones.

Strong girl. Because survival, after all, is just a state of mind.

Then she turned to me, and through a mouthful of mouse added, 'But boy, you're one twisted mother.'

I smiled back. I figured I'd been called worse.

43

RUNNING WILD WINS

IT TAKES COURAGE and faith and trust to put your life in a stranger's hands. And it is a privilege I will never take for granted.

Whether it is Kate Hudson screaming at me while swinging out of a mountain gondola or lost deep inside a narrow mountain tunnel in the Italian Dolomites, as pigeons flap around our heads in the pitch black. Or Jesse Tyler Ferguson dropping avalanche bombs from a helicopter and having to hold one in his hand as the fuse and the seconds tick away precariously before I instruct him to throw it. It is all about trust and faith.

Or Drew Brees, the legendary NFL quarterback, and me wrestling a croc in the Panama jungle and me smashing a rock down on the knife handle he was holding. Millimetres between hitting the mark in the moment, or crushing his hand that was insured for over $30 million. Split-second decisions, but ones that build friendships and trust that can last a lifetime.

It is like a drug to me. Creating bonds in the big moments and coming through them – exhausted, adrenalized, but united. I love that. And so do audiences. It was why over the first four full seasons of *Running Wild* for NBC we made the show a huge hit.

Inevitably, Discovery saw some of this success we were having and came back to the table, wanting to partner on some fresh new projects. I wasn't going to argue.

It was quietly satisfying for Del and me to see this happen, especially after all the confrontation we had endured before. It was all water under the bridge, but having them come back now, wanting new formats and shows from us, was sweet.

They might have lost out on *Running Wild* to a primetime network in the US, but instead they now agreed to buy the shows from NBC and us, in order to air them as what is called 'second window'. This was a good result. A double payday for the same amount of work, and Discovery still got some proven shows with big stars for their international territories that wouldn't have seen the originals in the US on NBC.

On top of this, we devised two new formats for them.

The first, *Escape from Hell*, retold some unknown but truly incredible human survival stories from around the world. We filmed all over the globe and I loved the show, mainly because it showed just how resilient and courageous human beings can be in the wild when under pressure. It was a life-affirming series for sure, even if at times quite gruesome. Plus it reduced the amount of time I had to be on screen. That part I liked especially, as our filming schedule got ever busier.

But the end result was that ratings simply weren't as strong as for *Man vs. Wild* and *Running Wild*, and the show only did one season.

Discovery rolled the dice again, and Del and I agreed to try and help them develop another new show. I really wanted to help them find an original format that would work as well for them as *Man vs. Wild* had, especially for all their international territories, where US-centred programming hadn't always translated very well.

Breaking Point was our next idea: taking people with chronic phobias (such as heights, snakes, spiders, rats, blood, confined places, you name it) and helping them overcome those fears through progressive exposure to them in the wild. The concept was great.

Again, we travelled the world and I got to stand alongside some amazingly brave individuals as I tried to help them through their fears. We actually experienced a solid 100 per cent success rate in dramatically reducing their phobias. All of our guests' lives were changed in some way for the better.

The positive psychological impact that the journeys had on those carrying such extreme fears was in large part down to the wild – and it was humbling to watch. The outdoors can be such a powerful teacher – and in terms of facing fears, it is hard to hide for long in the wild. If you want to get over any of those sorts of phobias then go on an adventure where invariably that 'something' will be around the next corner. Face it. It's the best chance you have of ever beating it.

Each of our contributors found an empowerment and healing that previous therapy hadn't been able to provide. It was special to see.

As Dan our camera operator said to me at the end: 'I'm not sure if you have either genuinely changed their lives and helped remove their chronic fears for good or if they were all simply too overwhelmed by the whole TV thing and your encouragement to actually tell you the truth . . . but either way, nice work. Job done!'

I'll take that.

Certainly everyone left smiling and happy.

But sadly, once more, the show just didn't rate as well as was needed, and despite all our best efforts it didn't get a second season.

Even though those initial new shows didn't quite reach the ratings Discovery had hoped for, they had brought us back together, and like a teenager returning home I was proud now to be back in the Discovery family, but under much better terms at last.

To this day, the relationship with Discovery Channel continues and is strong. They will often buy our other NBC, ITV and Channel 4 shows, and then air them worldwide after their initial release. Del and I like it because it keeps us tapped into the global viewership that the Discovery brand and platform always brings.

The important part of all this for me is the respect both of us had found for each other. There is no doubt that Discovery had helped give birth to my TV career, and with hindsight I saw it simply like this: I was the petulant teenager, Discovery the slightly heavy-handed parents; we'd had a bust-up, I'd stormed off, they'd then tried to cut me off; I'd tried to go it alone, failed a few times, but had then finally got going on my own two feet. And when I came home it was all tea and biscuits and niceness.

My mum had always taught me to avoid, at all costs, big quarrels in life. They waste energy and rarely lead anywhere good. She was right. Being friends again with Discovery was much better for everyone.

I will still always try to pop in and visit my 'parents' back at Discovery HQ if I am passing their London or Washington DC offices, and I will always regard both Discovery and especially David Zaslav, the ultimate boss and CEO there, as both friends and family.

Our brands have always been naturally intertwined, standing for adventure and empowerment. That's why when we did other new projects with Discovery, such as the *Survivor Games* series we launched in China, we made sure the show was co-produced by Bear Grylls Ventures (BGV) and Discovery Studios. A partnership of respect, where everyone wins.

Although running multi-million-dollar productions on the biggest platform in the world, terrestrial Chinese TV, isn't always straightforward – as I was soon to find out.

44

BREAKING CHINA

FILMING AN ORIGINAL show for Chinese TV with Chinese stars had been a clear goal of mine for a good few years during early *Running Wild* seasons.

Discovery Channel had previously launched into China, and the main feature series they'd led with was *Man vs. Wild*. As far as the government was concerned, *Man vs. Wild* was considered neutral, positive and aspirational to a new, young generation of Chinese. I took that as a great compliment.

It was also a clear moment that cemented brand BG into the world's biggest country.

On the back of the success of *Man vs. Wild* in China we were approached to develop and film a new primetime format for Dragon TV. We planned to film it in the heart of the Chinese jungle, and we called it, simply, *Survivor Games*.

To be able to return to that incredible country and film this new show was always going to be special, and we were set to take some of the best-loved Chinese stars as one large group on the adventure of their lives.

None of us really knew what to expect.

Just getting to the location was an epic (and I have known a few horror travel journeys), but whatever I went through, the crew had it ten times worse. By the time I arrived, all the team were pretty broken in terms of exhaustion, not from the physicality of the journey or the preparations but mainly from the endless late-night Chinese production meetings that just went round and round in circles. The end result was always the same. The next morning, even more Chinese execs would turn up ready for action.

If we ever thought corralling our own small team around a foreign country was tough, then try it with 150 Chinese crew, few of whom spoke a word of English, but all of whom carried a brand-new camera on their shoulder. We knew this was going to be an adventure.

Once we got beyond all that stuff and actually got on to the ground and into the jungle with these celebrities, it was an incredible time. It rained non-stop, and we had some long old days, but the thing I remember best is the way the Chinese stars started to warm towards values like selflessness and humility. Not the sort of values you might believe hold much sway when it comes to survival. But in the wild, selflessness and humility are top of the tree. If you're a big ego, if you're selfish, or you consider yourself too brilliant, you tend to end up abandoned, eaten or killed.

The celebrities each had to overcome some big fears and obstacles, whether it was heights, spiders, caves or white water, and it was powerful to be in the thick of the emotion with them. There is something beautifully universal about both courage and fear, and how people manage them. And those moments always bring people close together. I really learnt to love many of those Chinese stars, against all my expectations. I maybe thought it was going to be just another trip, another job. But so often, being in the wild and facing challenges together somehow unites people.

I will never forget one time when I had arranged for them all to eat some boiled bulls' testicles (as you do). We were all huddled on the side of a raging river in the rain, and Stani Groeneweg, one of our safety team, passed me the box of balls that we had sourced from some farm on the outskirts of the jungle.

As I opened the box, I took a sharp intake of breath at both the size of these things and the stench. I knew at once that these testicles were going to be hard work to swallow.

Determined always to do anything that I ask others to do, I grabbed a pair and held them aloft.

There was a brief moment of stunned silence followed by a loud outpouring of uncontrollable laughter from every single one of us at the sheer ridiculousness of it all. We had no need for our team of translators right now. Some things are just wonderful without words.

It remains my favourite photograph from all of my time filming in the wild, all ten of us holding those balls aloft, rain pouring down around us, and every single one of our faces crinkled with hysterical laughter.

In many ways that moment encapsulated everything that I love about survival. No doubt it was gross (the balls exploded as we bit into them), and undeniably funny (to see the individual horror produced by said explosions), but it was also magnificently unifying in its humour. What all of us as a crew felt (and they were all creased over as well, trying to stifle their laughter) was a great love for the beautiful spirit of those Chinese celebrities, and for the way the wild knocks our rough edges off and then puts us back together better, stronger, humbler. We become close to each other in nature in a way that is rarely found in normal life. It is like the natural world knows best. And when done right, the wild unites people of all cultures, languages, religions and nations through shared moments of hardship, endeavour and wonder.

One evening, the Chinese celebs exacted their revenge and tricked me into a dare that resulted in me agreeing to being strung up by my feet off the arch of an abandoned railway bridge. They then sat around below me drinking some Chinese liquor and toasting my good health. I admired them all the more for their boldness. I guess I had it coming.

The series ended spectacularly when we reconvened back in

Shanghai and were all invited on to the biggest chat show in China. I did a big 400ft rappel down the outside of the skyscraper that was home to the Chinese state broadcaster, before running into the studio live on air to meet up with the celebs for the first time since we had parted ways in the show's jungle finale.

The interview went well until they brought out a giant cake to celebrate one of the celebs' birthdays. As we huddled around the table, one of the guys pinched me. I discreetly put a little bit of chocolate cake down the back of his neck in retaliation, and before we knew it the whole interview had descended into one spectacular food fight live on air.

The Chinese host was horrified and eventually they had to cut filming, with all of us falling about laughing, scrapping, covered in cake and having fun. Once again, the wild had built those friendships. I think the channel somehow managed to cut all those antics out of the interview, sadly, but for me it was the perfect end to a great first season of a new show for us in China.

Somehow I seem to have picked out just the funny parts, and I don't want to give the impression it was all easy for our team, as the truth is it was a tougher shoot in terms of travel, communications, culture, logistics and local crews than any other show we had ever done before. But through it all I developed an enduring respect for the strength of character and relentless tenacity that are at the very heart of Chinese culture.

The *Survivor Games* series finally aired to some huge numbers, and on the back of this we decided to simplify and refine the format ever closer to *Running Wild*. We knew that *Running Wild* was now a well-established hit in other territories, and it felt smart to unify the two shows and bring the one-on-one concept to China. We renamed it *Absolute Wild*, and for this second season we went out to some of the best-known Chinese pop stars, sports stars, film stars and business moguls in the country. The response was incredible.

For me there are a few standout moments from the many episodes of *Absolute Wild* that we did. First, holding the end of a rope, trying and almost failing to control Yao Ming as I lowered him off a cliff face. The rope went bar tight under the tension and one of the crew had to dive in and help me to hold him up. At 7ft 6in, Yao Ming was the tallest active basketball player in the NBA. The pair of us climbing, hiking and camping together was like watching deleted scenes from *Gulliver's Travels*. He really is a giant of a man, but he also has a huge heart, and a passion for wildlife conservation, which is a love that we both share.

In fact, Yao and I have worked together since, with the Tusk Trust, helping spread the message that rhino horn and elephant tusk poaching must stop if we are to protect Africa's wildlife for the future. What I learnt is that when Yao speaks, many listen.

Which reminds me of the time Yao and I tried to 'break out' from the back of a police car (after we had persuaded the police to do some filming with us). Poor Yao was crammed into the boot and could hardly move. He looked at me and said, 'Bear, we've got to get out of here somehow.' In the end we managed to use my bootlaces to friction-burn through our plastic cuffs and we busted out of the car while the 'cops' were acting distracted.

Then there was the moment when I ended up swinging on the end of a rope, 1,000ft above a huge mountain gorge, suspended 80ft under a helicopter with the Chinese Olympic swimmer Fu Yuanhui. At possibly the worst moment, and for no apparent reason that we could decipher during the investigation we did afterwards, the electronic helicopter rope attachment point malfunctioned. In the blink of an eye the rope released itself. It was only the thin back-up rope that Stani had tied around the seat belt attachment point at the last minute that saved our lives.

Or stripping down to our undies with Robin Li, one of the wealthiest Chinese billionaires, in the middle of the Chinese high-altitude plains after almost getting stuck in quick mud. It was freezing cold and in the end we made two improvised 'cloaks' out of the bleeding fur skin from the carcass of a rotting yak. We looked quite a sight hiking across the plateau.

These moments come and go in the blink of an eye. But they are ones that I will never forget. Sometimes it seems so surreal when I think of all these crazy situations I have found myself in. A big old haze of adventures, friendships and close shaves. But when it comes to those adventures in China, I couldn't be more grateful to have been embraced so warmly by so many Chinese fans.

And I had a feeling that it was another new beginning.*

---

* As I write this, some five years on from those early Chinese shows, we have just broken ground on our first Bear Grylls Adventure outdoor theme park in China. Being built on a scale that I find hard to understand, our Chinese partners are creating 'mountains' and 'river rapids' in order to make adventure fun, safe and accessible to millions of Chinese. This is planned to be the first of many BG theme parks in the country. It's like the spirit of adventure transcends language and borders. And in this regard, we're only just getting started.

45

NEVER STOP EXPLORING

AROUND THIS TIME, I was getting more and more of an urge to do something that was rooted in adventure but was firmly away from the TV cameras.

Nowadays, the idea of going on an adventure and not filming it is an almost impossible concept for many people to fathom. But I had got into adventure as a kid as an escape from school, I had joined the military as an escape from university, and I had started doing expeditions as an escape from the constraints of a job. The one thing I had never got into adventure for was to be on TV.

Fast-forward a few years, and almost before I knew it my whole life had become overrun with TV productions and camera crews and making shows – week in, week out. I loved the adventures, always – don't get me wrong – but the filming side of things has never been either easy or fun for me.

Even when Gilo, one of my best buddies, and I came up with the idea of flying powered paragliders over Mount Everest, somehow TV cameras became a big part of it all. We had a crew from Channel 4 and another from Discovery following us around, and I was back into filming and work mode. Again.

Despite this, Gilo and I pulled off the mission by the skin of our teeth. Gilo was the real hero of that one, designing and building a paramotor engine that could do the impossible and fly so high into rarefied air.

Together, we eventually broke the world altitude paramotor record, on our way towards the highest peak on Earth, but his machine died and began rapidly descending just before he managed to get above Everest. In a few minutes he was but a speck, thousands of feet beneath me.

Suddenly I found myself on my own up there, and really pretty scared. But there is something about just being in a situation and having to deal with it that focuses the mind. And as we all know, panic rarely improves anything. So I tried to keep calm, and I kept ascending at full power, hoping for a miracle.

Gilo's amazing machine, which was a truly remarkable feat of engineering in itself (and which played a part in earning him an MBE from the Queen a year or so later), did the rest. I am in no doubt that Gilo's hard work and genius allowed me to soar to over 29,500ft, just above the highest point on Earth – and all under a little parachute with an engine strapped to my back.

Looking down on and across at Everest was a special moment. Not least for the fact that for the first time in weeks there was no director asking endless questions with a camera in my face. Up there I was very alone.

As my buddy the legendary US BASE jumper and freeride skier J. T. Holmes said to me afterwards, 'For me it was when you suddenly found yourself solo, after Gilo's machine was unable to climb any more and had to turn back to base camp. Cold, not much air, compromised communications . . . it would have been real easy to head for the nearest reasonably safe landing area choice, rather than squeeze that throttle and go higher. It came down to a choice, like [American mountaineer] Conrad Anker on Meru Peak: do you go up or down?'

It was a choice I don't regret, but there is no doubt I got a little lucky. And in my mind, as I said, the real accolade must go to Gilo. Either way, it felt like a long way to go to get some peace and quiet.

At the end of the day, Gilo and I had, above all, wanted to go and have a huge adventure – just for fun. And that sense of fun, friendship and adventure has always been more important to me than any Guinness World Record or TV moment. I vowed that the next big expedition, if we did one, would be strictly camera-free. A chance to remind myself how it all began.

Rigid inflatable boats and the ocean have always been such a big part of my life. Ever since growing up on the Isle of Wight and fixing up an old wooden boat with an improvised steering wheel attached to a 2hp outboard, I had been hooked on the ocean from a very young age. And the bigger the sea the better.

When I taught sailing as a teenager back home in the summertime, the best days were the ones when a gale would blow up and one or two of us were sent out in the small sailing club RIBs to try and rescue all the dinghies caught out in the squall. Towing in ten dinghies with the rescue RIB through the driving wind, rain and waves, with a trail of terrified kids behind, was a happy place for me.

I had led a big Arctic Ocean expedition in the year our eldest son Jesse was born, and the year before my TV career started. We were a small team of five, and we had crossed the North Atlantic, just below the Arctic Circle, in an open RIB. The attempt had taken all of us on that small boat to the edge when we got caught out in several huge storms hundreds of miles off the coast of Greenland. We were incredibly lucky to survive that 3,000-mile journey and I had promised to Shara that I would never take on a trip with that level of risk again.

She had gone through the horror of having the Royal Navy, who were one of our sponsors for the expedition, flag up that there was a significant chance our craft had gone over and been lost in one particularly grim Arctic gale we had got caught out in. Our tracking device had gone down, and the Royal Navy and our ground team had the Icelandic search and rescue on standby to come and try to find us. Luckily we made it through the storm and back into comms with minutes to spare before the search operation was launched.

The experience had beaten all sense of bravado out of all of us, and I certainly wasn't in a hurry to get into another expedition on that scale and with that level of danger. But maybe taking a RIB for a few thousand miles through the infamous North West Passage, the other side of the Arctic to where we had been before (albeit much further north), might be a great escape from the world of TV . . .

Just a small trip, I figured. A fun one. That always seems to be how so many great adventures begin.

We raised some funds, secured a boat and prepared for the trip, all in between filming new seasons of shows. 'A little summer break from filming' was how I sold it to our production team, and to my family.

It actually turned into an incredible trip. It is not often in the modern day that you find yourself in territory that is genuinely uncharted. The only other time we had experienced this had been a disastrously over-ambitious expedition I led to try and fly powered paragliders on to the top of the Angel Falls, the highest waterfall on Earth, hidden in the Lost World among the remote table-top mountain tepuis of Venezuela. We had failed spectacularly.

In fact, it had been nothing short of miraculous that we hadn't got sucked up into some of the giant tropical cumulonimbus storm clouds that hugged the vast jungle cliffs of the area. With no comms, no maps, no supplies and no clue, we would have been toast.

The only safety gear I remember having was a roll of paracord in case I crashed into the high canopy of trees, so I could abseil down safely. The mind boggles at our naivety. (Especially when we realized after the trip that paracord breaking strain is about 50kg.) But there is also a part of me that kind of loves the ambition. After all, adventure only happens when things go wrong, as they say.

Thank God we failed though, and we got spat out of those aggressive torrential rain clouds, eventually landing in a small jungle clearing near the foot of one of the tepuis. If we had managed to get up to the falls themselves, we would still be there to this day, of that I am certain.

But the whole trip had been amazing. Even getting there was an epic experience, not to mention the camping, flying and living in such a wild place, where the most modern maps of the time had a giant white blob in the middle with the word UNCHARTED stamped boldly across it.

That to me was the best part. Today, all the world has been fully satellite-coded and covered, but back then Google Earth didn't exist, and there was a certain magic to heading off into the unknown.

46

ICE CALLING

THE ARCTIC'S NORTH West Passage was also appealing on other levels, apart from there being no cameras. We would be off grid, off the official maps, and there was also the fact that a complete navigation through the entire ice passage had never been done before in a RIB.

Ran Fiennes had done part of it as a section of his epic Transglobe Expedition three decades earlier, a feat I vividly remember reading about as a kid. What a legend of a man Ran is. He has always been such an inspiration to me, and a kind friend ever since I had my first spate of negative press during those early *Man vs. Wild* years. He had so thoughtfully written me a postcard that simply said, 'Ignore the buggers. You're doing OK.' I held him in even higher regard after that.

The final appeal of the North West Passage was the harrowing story of the Royal Navy ghost ships HMS *Terror* and *Erebus*, which went missing in 1845. The ships, under the command of Sir John Franklin, had been the pride of the British Empire and the unconquerable Royal Navy. Their mission was to probe and discover a navigable polar shortcut through the North West Passage. This would provide an invaluable trade route for the British Empire around the world that would save months on the conventional route around Cape Horn in South America.

The problem was that both ships vanished, and no trace of them had ever been found. The early explorer John Rae discovered some years later that local Inuits had provided a seal to some starving Europeans only to return to find thirty human corpses. Rae discovered they were the bodies of the *Terror* crew. In a chilling letter to the Royal Navy Admiralty he wrote: 'From the mutilated state of many of the corpses, and contents of the kettles, it is evident that our wretched countrymen had been driven to the last resource – cannibalism.'

It was a grim but captivating story. And despite numerous expeditions over the last century, no one had ever found any further sign of either HMS *Terror* or HMS *Erebus*. Their last known position was in the Victoria Strait, in the high Arctic.

Cannibalism aside, what I didn't want to get into was another long expedition of the like we had experienced on our previous Arctic RIB crossing, which had become a masterclass in coldness and misery while facing potentially imminent death. We had done that. This time, I simply wanted an adventure with great friends, to get off grid, away from the cameras – and to do something that hadn't been done before would be a cool bonus.

One thing that was becoming clear was that this trip would be much safer if we could somehow enlist a support vessel that had some ice-breaking capabilities – just in case. I had learnt my lesson. The costs were huge, but I knew it would make the difference between a fun trip and a potential horror show, so I persevered to find one.

The clincher came when we did a short overnight sea trial in summer in the UK, to one of the Channel Islands and back, with our sponsors. We got caught in a horrible storm with some big old seas and, in our small RIB, it quickly became quite focusing – as anyone who knows what the currents around those islands can be like will attest, particularly when angry and at spring tides.

By the time we pulled into a little harbour for shelter, some eight hours later, we were pretty exhausted. Tim Levy, who was our lead sponsor and CEO of a big company in London (and who had been really keen to join us for the Arctic journey as well), was pretty shaken up. I knew it would be the perfect time to suggest the support vessel.

'But it will be expensive, Tim, that's the only thing. I reckon we can do without it, and if we get caught in a few storms like this one, I am sure we will be fine.'

I waited for his response.

'Bear . . . whatever the cost . . . we just cannot go to the Arctic with no back-up. Whatever the cost, we've got to get a mother ship.'

Job done. The formidable English Channel can be a persuasive place.

And in hindsight it was genius. The ice-breaking vessel that we eventually found made the entire journey much more feasible. We could store supplies and fuel on board, and on more than one occasion we moored up alongside the vessel when a storm hit. And the warm food and hot showers were sweet relief. Once bitten by the Arctic, twice shy.

47

THE ULTIMATE DISCOVERY

EARLY ON IN the North West Passage expedition, our mother vessel found herself approaching thick ice. They would either need to wait for the weather to change or the ice in the Passage to ease.

The captain wanted to wait and see if the weather would break up the ice further over the next few days.

'What about if we take the RIB and try to find another more ice-free route?' I suggested. 'Maybe we head two hundred miles south-west, through all this area, and then we meet you in a few days at Cambridge Bay,' I added, pointing at the charts.

'You'll be in uncharted water there,' the Norwegian captain replied. 'You'll have no idea of depths or hidden rocks or any number of dangers.'

Dave Segel, who was also one of our crew, tapped my foot, raised his eyebrows and smiled. (Dave had burnt out most of his adrenal glands when he was a young City guy hammering the hours at every which end. This lack of healthy adrenalin and fear sometimes made him a dangerous person to hang out with.)

Historically, any ice-breakers going through the North West Passage followed the exact same route Roald Amundsen had discovered a hundred years earlier. (Amundsen had previously become the first man to reach the South Pole, beating Captain Scott, who had died on that fateful expedition in the Antarctic.) This exact, narrow Arctic route through the Passage had since been traversed by US, Canadian and Russian ice-breakers and a handful of other smaller boats, but not many.

Today, with global warming, the route opens up almost continuously in the summer months, but for most of the previous decades the number of vessels through the Passage had been remarkably few.

Once you deviated off this main charted route, though, the Garmin plotter came up with the wording NO DATA AVAILABLE.

I knew instantly we would be going that way.

We had an incredible time navigating at high speed in our open RIB through the myriad of islands that dot this part of the sea, all of which were of low elevation and covered in shingle, snow and ice. It felt amazing. Although there were a bunch of times when we had some very narrow misses with rocks and unseen shingle banks and ice floes.

At the end of one day, we pulled in behind one of these islands to shelter from the biting Arctic wind and to find a spot where we could anchor, get ashore and put the tents up.

We picked the biggest island we could spot with the best natural anchorage, bearing in mind none of the islands rose more than about 20ft above sea level. By the time we got the tents up and the stoves on to melt ice, it was getting as dark as it ever does in the Arctic summer months: a faint glow to the sky as if the world is locked in a permanent state of dusk. An awesome sight.

Ben Jones, our Welsh engineer – who had never left the UK before, and for whom we'd had to sort a passport before the trip – wandered off for a smoke.

'Holy crap – what's this?' he shouted, looking down at a human skull he had accidentally kicked.

We soon discovered that the little island was a treasure trove of rough-and-ready Western-looking graves, human bones and skulls, rifle cartridges, and one particularly long length of thick wood, half submerged in the shingle, that resembled an old-style spar or mast. We also found a large area of blackened rock, as if the mother of all signal fires had once burnt in this place.

Dressed in huge yellow dry suits, balaclavas pulled down against the chill, Dave with tape strapped across his face to protect his busted nose which he had broken when he hit his head during a storm we'd gone through a few days earlier, we looked a real sight. We stood there silent, Jonesy staring down at the skull now in his hands.

Two things were certain: one was that no Western boat would have passed through this part of the Victoria Strait since the mid-nineteenth century, as we were well west of the conventional North West Passage navigable route, and boats don't go where there are no charts, unless they are small, shallow-draught and nimble like our RIB. And RIBs in the Arctic, this far from settlements, are truly few and far between.

Secondly, this whole scene didn't look Inuit at all. And who would be out here burying bodies and burning huge fires anyway?

We looked at each other. And we all thought the same thing.

Franklin.

The conclusions were clear to us, although we tried not to get too high on discovery before we could carefully analyse the facts. But from what we saw, we had stumbled across Western graves and potential signs of the desperate final weeks or months of dying men, waiting, hoping, praying for rescue. They had burnt a serious amount of timber that must have come from somewhere (bear in mind there isn't natural forest for a hundred miles up there), and there was no other clear reason for anyone to be dying or burying each other in substantial numbers this far away from civilization yet so close to the last known spot of both *Terror* and *Erebus* . . .

We got pretty excited.

When we eventually made it home, we reported all this to the Canadian authorities and marked the exact spot for them. In the years since they have found the wreckage of both HMS *Terror* and HMS *Erebus* supposedly near to King William Island, just to the south of where we were, but they never publicly released the exact locations in order to preserve the wrecks from bounty hunters – or so they said.

I like to think we accidentally stumbled upon the last place where the majority of the *Terror* or *Erebus* crew had waited while consuming their dwindling supplies and hoping for rescue.

I also believe that if I knew that rescue wasn't coming and my ship was frozen solid for winter, trapped in the ice, and supplies were running out fast, then I would start to burn parts of the ship to stay warm and to at least keep a signal fire blazing, just in case.

I mean, if I was going to starve to death, and was busy burying crew members left, right and centre, I'd want to at least die warm . . .

But what do I know?

48

TOP SECRET

THE OTHER MOMENT I will always remember from that expedition happened at the last place we camped, in a tiny inlet in the Beaufort Sea called Pearce Point.

The Beaufort Sea is a wild and remote part of the planet, at the western end of the North West Passage, and it lends its name to the Beaufort scale of wind speeds and sea states. It is fitting really. I have been to many places around the world, but the Beaufort Sea to me has always felt closest to the true ends of the Earth. More so even than Antarctica, or the jungles of the Lost World, or outer Siberia. There was something truly remote, cold, hostile and frightening about it. Huge, unpredictable waves, and dark brown water, as if the wild seas had frothed the shingle into a dirty mush for hundreds of miles in every direction.

Get in trouble here and no one is coming to get you. It has that ominous sense of foreboding.

By the time we swung into the shelter of Pearce Point we were done. The battering of the sea and the long hours had taken their toll, but the team had been amazing, and even Tim had endured stoically. He was the first sponsor we had ever really taken on a trip, and I would take him again in a heartbeat. Humble, kind, fun, strong. And in turn, he'd had the trip of a lifetime.

This was now our last night before flying south from a small Inuit airfield.

As we pulled into the cove we spotted a large polar bear loping across the shore line. Awe-inspiring to see, so huge, and undoubtedly a predator not to be underestimated. I already knew not to mess with polar bears. They are the ultimate power hunter, even able to catch whales when desperate. Only a few years earlier some English students on a field trip to the Arctic had been stalked by a polar bear. One of the students lost his life, his head chewed to bits. They eventually tracked down the bear and discovered it had been suffering from a tooth abscess. The inquest concluded that the bear had simply been using the boy's head to ease its toothache.

Choose your battles in the wild, and never get complacent. Another mantra of mine.

And don't screw with polar bears.

We watched the bear look at us, then it slowly moved away as it saw us approach the shore.

We would need to keep an eye on it.

I have learnt on expeditions always to play to people's strengths. Tim had proved himself amazing at figuring out some really pretty complex mathematical equations in order to monitor our fuel situation during the trip. How long we could travel for, at what speed, against what sea state, at what fuel burn rate. It was a vital task to get right: across such big distances between Inuit ports or to our resupply vessel, our margins weren't always that great.

'Keep going at eighteen knots, then at three p.m. up it to twenty-two knots, and as long as we don't hit any more of the pack ice we should be good until our next resupply,' Tim would shout over the noise of the engines and wind.

Amazing skills. No wonder he had made such a fortune in business.

But when it came to setting up camp, he was maybe a little less helpful. Play to people's strengths.

'Hey, Tim, if we set up camp, will you do a short perimeter patrol of the bay to make sure that polar bear has moved right away from here?' I suggested. 'But not far. Just keep an eye out for us.'

I passed him the shotgun, but intentionally with no cartridges. In truth, I could see the bear was gone and I was just giving Tim some space while we set up camp.

Fifteen minutes later, the tents were up and I looked around, wondering where Tim had gone. There was no sign of him. Where the hell was he?

Eventually we spotted him on the far horizon, shotgun in hand, jogging along a ridge line with the polar bear moving steadily away, some 500 metres in front of him.

The problem was that Tim didn't realize the shotgun wasn't loaded.

We all sprinted off towards him, madly waving and shouting to him to come back.

It would have been the ultimate dead man's click if, God forbid, that polar bear had charged at Tim. But, luckily, all was fine and we never saw the bear again.

The other final, cool thing we discovered up in the Arctic, tucked away down the end of a very remote inlet, in shallow hidden terrain, accessible only by small craft, was an empty but clearly functioning military base station. The place consisted of a cluster of ice- and weather-proof buildings, marked pathways, monitoring equipment, aerials, radars and antennae, all carefully arranged and connected. The buildings and bunkers were made of corrugated shiny metal, all dome-shaped and sunk into the permafrost. It was apparent that the place was well maintained, although it looked unmanned.

We had a good look around and then got the fear we were being watched, so didn't hang around long. The many CCTV cameras, KEEP OUT and THIS STATION IS MONITORED 24/7 signs did their job.

How often it got checked, who knows? And we never did discover its purpose. But it was amazing to see. No doubt the US and Canadian military keep many covert places like this in remote parts of the Arctic and beyond, for various unsavoury reasons . . . just in case the Russians ever stop playing ball.

And then just like that, before I could blink, we were all back south, back to normal life. I was straight back into filming, and true discovery and exploration – historic wrecks, secret bases and polar bears – quickly seemed like an age ago.

Life is strange like that. Too fast sometimes. But it had been a great trip and the perfect antidote to the constant cameras, and the pressure of having always to be on.

49

POTUS CALLING

IT'S AUGUST AND we are at home on the island, finally enjoying some time out after a crazy first half of the year. I totalled that we had been filming in about twenty different countries around the world in the last few months, and I was beat. Our precious family time together on the island had been the light at the end of the tunnel for me. It always is.

My phone rings at midnight one evening. It's Del.

'You sitting down?'

Del then tells me the White House have reached out to our production team to say that President Obama is a fan of *Running Wild*, and would we consider shooting an episode with him in Alaska? He was heading there for an official visit and the ask was to show the President some of the dramatic effects of climate change close up. He wanted to see glacial erosion for himself and had asked for me to guide him.

The time frame meant that we would have only two weeks to prep the journey with the Secret Service team ahead of his visit. But the team were already on it, and NBC had just agreed to air it as a *Running Wild Presidential Special*.

I remember clearly the slight look of disappointment on Shara's face that evening. No doubt it was a great honour and an invitation we really shouldn't decline, but she also knew that our precious family time together on the island, which we protect so fiercely, and for which we normally turn down every invitation and offer, was going to be broken. But we both knew it was the right thing to do in this instance.

Although I do remember Shara suggesting, in a way only she could, without a hint of irony, that maybe the President could come instead to North Wales to shoot the episode. Sadly, that wasn't going to fly.

The real problem we had, though, was that most of our safety team was in China, planning journeys over there for our autumn filming schedule. Instead, we asked my trusted friend Dave Pearce to go out from BGV to work with the Secret Service to plan a route and liaise with them on all the security implications that surround the President of the United States (POTUS for short) when he goes anywhere.

I had no idea quite how crazy that process would be.

A week later, I rang Dave for an update. I mainly wanted to know how many Secret Service agents Dave reckoned would be with us on the ground on the day. As ever, I knew that a small crew was always the key to being able to make a good show and complete a fun journey. I hoped it wouldn't be more than a handful of security detail. After all, we would be in the middle of nowhere.

Dave replied simply: 'BG, it's going to be Marine One plus three other choppers to drop him off. Full air cover, roadblocks for the motorcade of twenty Secret Service vehicles to bring him to the journey start point. Then about fifty or sixty agents, minimum, with us, hiking the journey and lining the route.'

'Right. So not quite the normal "leave your entourage and come on your own and trust us" model?' I replied.

'No, Bear. Not this time.'

Forty-eight hours from when we were due to start filming with Obama, I flew from our little Welsh island by chopper to Heathrow airport and got straight on to a flight to Chicago. I had three very tight flight connections in order to reach the remote corner of Alaska where we would be meeting the President.

I had chosen to fly out way too last-minute. Part of it was that I didn't want to leave the family alone on the island for long, especially when it's bad weather, and part of me still felt a little guilty about leaving them at all during our precious family time.

Retrospectively it was a really stupid decision to cut it so fine. *It's the President, you idiot, leave yourself enough time.*

Once we boarded our flight, an announcement flagged that there was an issue with the loos and that we were looking at a two-hour delay. It was two hours I didn't have to spare. The maintenance men on the plane who were trying to repair a fuse in the bathroom didn't seem to be in much of a hurry. I even ended up helping them at one point to the amusement of the cabin crew. They finally got it fixed and we were off.

British Airways' special services team were amazing from then on and met and escorted me between planes in Chicago, and by the skin of my teeth I collapsed on to the final plane that would bring me to location.

It was almost 2 a.m. by the time I arrived in the crew motel on the edge of Kenai Fjords National Park. We were due on location at 7 a.m. by chopper, to allow me to quickly scout the area before airspace would be closed in preparation for the President's arrival with us at 11 a.m.

It had all been way too tight. Another lesson learnt. Don't be a rushing idiot when it comes to the big stuff.

I couldn't sleep those final few hours. Adrenalin, nerves, jet lag – plus batting off endless emails of suggested questions that NBC producers back in LA wanted me to ask Obama. I ignored most of them. I had long since decided that I would just do what we always did as a team. Keep it honest (mistakes and all), keep it fun (if in doubt revert to toilet humour), take a few calculated risks (even if the

Secret Service had previously said no, push it a little), and chat about the stuff that comes from the heart rather than anything political (struggles, high points, family and aspirations).

By 10.30 a.m., finally, we were ready.

The scouting was complete, the plans with the Secret Service all squared away, weather was a go, and there I was stood on that Alaskan river bank, alone, waiting.

Three military helicopters swept the area one last time. I noted the spots where the Secret Service snipers would be in position by now. As planned.

President Obama was due any minute.

I calmed myself.

Life had somehow, strangely, led me to this moment. And I knew I could do this.

Just breathe, be calm, be polite and have fun . . .

OK, there he is, let's do this, it's time to go to work.

50

BEAR KILL AND BEAR SPRAY

THE LEAD SECRET Service agent had flagged the anti-bear spray that I was carrying. He'd told me in no uncertain terms that my suggestion of showing the President how the deterrent works was a bad idea.

But some rules in life are meant to be bent ever so slightly.

It would be how I would break the ice with a normal *Running Wild* guest – do something fun. It generally works beautifully, and then the journey begins.

I told the President that we were firmly now in grizzly country and that we should be careful – and that you should always carry some anti-bear spray as a deterrent. These canisters are powerful tools that can spray up to 25ft at a charging bear – and if used right, they should deter an attack, at least temporarily, in order to give you a chance to escape.

I always find it kind of tempting to give things like that a quick test blast. I guess it's the same temptation I always feel to give a fire extinguisher a little squeeze.

I said to the President that maybe we should give it a quick go, to test it before we went into the forest. He smiled and said, 'Sure.' Then he looked me up and down and added, 'By the way, are you always so muddy?'

I smiled back. I had intentionally not washed my gear from our last expedition. I thought it would help keep our journey together feeling fun and informal.

'Um, well, yes, generally. I like grubby clothes. They have history.' He smiled back.

'Anyway, Mr President, grab this, hold the can facing forward, and blast it downwind . . . Oh, and watch out for the Secret Service guys,' I added with a wink.

The lead Secret Service guy gave me a resigned shake of his head.

The President gave the canister a blast downwind, we both chuckled, and then, along with a whole mini-crowd of aides, assistants and close protection agents, we set off together towards the glacier.

Within a few moments of leaving, the wind changed and I suddenly got a waft of powerful anti-bear spray in my eyes. Immediately tears started streaming down my face.

The Secret Service guy looked at me, smiling.

Now who was the idiot.

The journey went by in a blur – and not because of the bear spray. But I remember so well a moment when the President and I were alone, waiting for the camera teams and the Secret Service to leapfrog ahead of us, as the trail had now narrowed to single file.

I'd just been telling him about how especially dangerous it can be if you disturb fornicating grizzlies – at which point the President had said he understood that one entirely. We both laughed, as anyone would, and then we stood there waiting for the camera and security teams ahead to be ready for us.

'It must be tough always having those guys around twenty-four/ seven,' I told him. 'I think it would drive me nuts.'

He smiled. He said how sometimes he wished he could just go and get a coffee like a normal guy.

'But surely you can just tell them to leave you be for a few minutes?' I said. 'After all, you're the boss of bosses.'

He laughed. 'Yep, it's a bit of a gilded prison to be honest at times, but it goes with the job. It's tough on the family, that side of things, for sure, though.'

For me, it was these little moments, when it was just the two of us, huddled in the bushes waiting for the teams to move ahead, that I look back on and loved the most. It's what the wild does: it levels people and removes airs and graces; it brushes aside the normal formalities of life. In those moments, the wild connects us.

I teased him about forgetting his phone code, and for not being able to take a selfie. 'I mean, is it really necessary to have a twelve-digit code?' I ribbed.

He nodded, then shouted over to one of the Secret Service team: 'Joe, what the hell's the passcode again?'

Joe looked decidedly reluctant to shout out the code from where he was, some 50 metres away. But the President pushed him.

'ALPHA-NINER-BRAVO-SIX-FOUR-TWO-MIKE-ECHO...'

It took four attempts for the President to punch the code in right. I felt pretty certain Joe was going to be changing the code later on that afternoon.

I had also been told that there was no way that the President would be eating the half-eaten salmon we had found the day before on the river bank. The remnants of a grizzly's lunch.

'No way,' the Secret Service had said.

Instead they had the President's chef prepare a salmon that they said they would 'swap in with the half-eaten one' that I had in my backpack, sandwiched between two clumps of moss to keep the fishy carcass together.

At the appropriate moment the chef would step in and we would switch the salmon out. That was the plan.

We had reached the snout of the glacier and I'd shown the President a few cool tricks about how to walk on slippery ice with his socks over the outside of his boots, and how to light a fire with a flint.

I produced the bear-kill salmon and slapped it down on a thin rock balanced over our little fire, and started cooking.

I could see the chef behind the camera, stood there, holding his silver platter covered with tinfoil, with the pre-cooked, pre-approved salmon waiting underneath.

He kept trying to get my attention, pointing at the platter, but I resisted. If the President wasn't going to insist we swapped it out then I wasn't going to suggest it either.

As it turned out, the President was totally game, and together we wolfed down the bear kill and he even shared a water flask with me – both protocols that would never happen in a normal setting, as I was repeatedly told afterwards by his team. I reckoned they had quietly loved the rule-bending spirit we brought to the whole adventure.

On our hike back we discussed global warming and the shocking extent to which glaciers are eroding, and the devastating effects that sea-level rising is having on poor coastal communities around the world. We spoke at length about how climate change is causing more famines, more extreme weather and more extreme temperatures. It was a poignant moment, where the most powerful man on Earth, who had so championed protecting the planet during his presidency, actually got to see the effects of global warming close up.

We discussed his hopes, his family, whether his kids ever got lost in the White House when they were little, if he ever pinched himself while carrying out his job, and many other vitally important national issues – none of which were on the NBC exec wishlist, but all of which normal people, me and our crew, were interested in.

At the end, the President turned to me and said it had been one of the best days of his presidency. He was out of the office, out of a suit and not being grilled on political issues. He said that he had laughed and been treated like a normal guy, and how he loved chatting about the regular things he really cared about, whether it was his kids, his wife, his hopes, his struggles or the planet.

We hugged, and then I said a short prayer for him. A few quiet words to ask for protection over him, for blessing on his family and his work, and for strength for the times ahead. The moment was never meant to be in the final show. But in the end, the prayer was included and it went out.

Many people digged it, a few criticized it. They always will. But some things just feel right to do in the moment; and the connection that the wild, good chats and a vulnerable deed such as praying for someone can create is often powerful and lasting.

Personally, I didn't regret it for a second.

As for the experience of taking the sitting President into the Alaskan wilderness, it's definitely up there. And a moment of pride for the family.

51

WHITE HOUSE LOCKDOWN

THE *RUNNING WILD Presidential Special* aired all around the world and on primetime on NBC. In the UK it got the highest ratings of the year for Channel 4 and I ended up winning *GQ*'s TV Personality of the Year as a result. That made my mum happy.

But above all, as a team, we knew we had done something special.

On the actual filming day in Alaska, as we were saying goodbye, the President had mentioned that I should bring my family to the White House some time. I took it as a polite gesture, nothing more. But as he left, one of his close aides took me aside and said he rarely ever said this, and that I should take him up on his offer.

Six months later we did it. I took the family to Washington DC over the Easter break, and before we knew it we found ourselves in the Oval Office with President Obama, who was dressed this time in a suit. I was still in trainers. I was actually way more nervous meeting the President in the White House than in the wild – somehow the wild always feels like home turf. To settle my nerves in moments like this I always remember my mum saying: whoever you are, every-one puts their trousers on one leg at a time.

Anyway, the President was so kind to Shara and the boys. We swapped stories in the Oval Office and laughed about silly family stuff. It was just us and him, hanging out for thirty minutes, and it felt like a real honour. The President then kindly signed three individual mini leather atlases for the boys, with the presidential seal emblazoned on them, and we said our goodbyes, parting ways to allow him to get on with his real job.

As we exited the Oval Office there seemed to be a sudden bustle of activity and we were asked to go in and wait in the anteroom. Something had just happened, our escort said, that they couldn't tell us about right now. They would be back for us in a few minutes. We looked at each other quizzically. A few moments later we were ushered down some stairs in the West Wing and taken along various corridors, with Marines hurrying to and fro past us.

Then suddenly we were passing the Situation Room. I recognized it instantly from news clips.

Soon we emerged from a side exit and made our way out of the White House, to many apologies from the staff about all the drama.

It turned out that, literally during the time we were in the Oval Office, there had been a shooting a mile away on Capitol Hill, and the White House had been rapidly sent into lockdown.

As we got back to the hotel, I received a flurry of worried texts from Del who had seen the breaking news online and knew we were there right at that time. I told him all was good. In fact, the boys had loved the adventure. Marines, lockdown and all. In fact, I'd hazard a guess that it had been their favourite bit.

Del simply replied, 'Classic BG. Adventure always seems to follow you around.'

As far as I am concerned, in life, in family and in work, I can ask for no more . . .

52

KRYPTONITE

IF YOU ASK our BGV team the one thing that is like kryptonite for me, they will all say it is doing interviews and long press days. It is why we work so hard to make sure I don't ever get tied into contracts guaranteeing a large amount of media contact. Instead, we keep the terminology loose.

I think it comes back again to having endured so many long press trips in my early TV years, sat in a stuffy studio doing back-to-back interviews for hours and hours. A morning of endless TV and radio shows across America, until I was given a break where I could run outside and breathe and close my eyes. Then back in again, but this time for four hours of international interviews. The same questions, the same stories. The same close-up cameras, that same sense of always being watched. Like a caged animal, it always made me feel really anxious.

It is why, still to this day, I struggle with doing press. I can feel my breathing start to get shallow and fast. It's called anxiety. A twitch. But it's also a survival response from somewhere deep inside. I know I am in a strange, uncomfortable situation. A hostile place that makes me hyper alert. Too much so. And I know where it comes from.

Ever since all that negative press around *Man vs. Wild* and Discovery Channel all those years ago, I guess I am a little like a wounded dog when it comes to interviews. At the time I had some tough interviews, and I look back and realize that I simply didn't have the skill or knowledge to deal with the questions very well. I was always on the back foot and often defensive in the face of what I saw as journalists just looking for headlines over facts. There were endless battles that I always felt like I lost, and I rarely did a good job of answering difficult questioning. Nowadays, I wince if I stumble upon some old interview. They were invariably pretty terrible, to be honest.

It is why now I just prefer to avoid press whenever I can. People are often surprised by this. 'But you're a celebrity, you must love the attention?' But for me, attention has never been the goal. Even as a kid, I just loved being quiet and on my own up some tree, listening to the pigeons or watching the changing sea from the beach. Nature was my safe place. It still is.

Press always felt like the opposite of that. I used to dread the interviews, and not being free to move around or just to stop for a moment and breathe. Everything about press, studios and interviews is about countdowns, protocols and performing. They are about being confined and controlled – three, two, one, you're now LIVE ON AIR. SMILE! I hate it. It wasn't just the fact that I would have to be answering the same questions over and over, it was more the introspection of being asked about myself all day long and being confined to a chair under studio lights that I struggled with.

Some celebrities love it. The questions and the chat. But for me it has always felt self-indulgent and self-serving. I have always preferred to let projects stand or fall on how good they are, or aren't. And I believe that great shows, like all great projects, always find their way, whether or not they are heavily promoted. We live in a world where everything tends to get oversold, then all too often under-delivers. I'd prefer to be known for the opposite.

The big lesson I have learnt, though, about doing press is simple. That humility is everything. That means it's OK to be nervous, to not know, to apologize, or to be quiet. It's OK to stumble on a reply or to take a moment. You don't have to be slick or loud or confident. TV is so revealing. It's always better to be honest and to be yourself, even if it is less dynamic or exciting than some producer might like. Being yourself is OK.

In addition to this, there are two things that have made doing interviews much easier for me.

First, whenever I can, I always throw the spotlight on to others and make them the focus. This is why I always try to do press that highlights something special that we are involved with, whether that be Scouts, the Royal Marines charity, TUSK conservation or our Be Military Fit outdoor fitness company. Make them the main attraction.

This is one of many reasons I love Scouting. And then I can divert attention on to the Scouts themselves. I will often take some Scouts with me on to talk shows. That way I can talk about them and the amazing things so many Scouts and volunteers have been doing, stuff that is invariably unsung and inspiring. Like any good family, they protect me as much as I champion them. All I have to do is throw the spotlight on to them. And that way, everyone wins.

Second is having something to show the journalist how to do, such as lighting a fire with chewing gum and a battery, or some self-defence move. It gets me up and moving and having fun again, and that, in turn, stops it feeling like an 'interview'. There have been so many occasions when I have used this tactic, whether it was disarming Jonathan Ross on his chat show when he tried to knife-attack me with a banana (as you do), or playing table tennis with Jimmy Fallon on his US talk show using two razor-sharp machetes as bats. A magician might call it a misdirect. But for me, it works so well.

I have had some memorable moments on chat shows over the years. Abseiling in from the roof for *The Oprah Winfrey Show*; pulling worms from my mouth while hanging upside down and feeding them to the comedian Dawn French; racing down the Thames with the Royal Marines in a high-speed boat before smashing through the huge frosted fake glass window front of ITV's *This Morning* show.

All fun moments, but moments with a purpose. To distract and divert. To keep the interview fast and fun. Just like that food fight in China. In. Out. Off.

Doing the *Piers Morgan's Life Stories* show for ITV was one of the most intense ones I ever did, with nowhere to hide, and I was so nervous beforehand I could hardly speak. I knew he would ask some tough questions and would go deep into some hard moments from my life.

Halfway through, he totally caught me off guard by showing some footage I had never seen before, of my late dad and me when I was a kid, the pair of us tinkering around in a little rowing boat. The emotion hit me like a wall, and I ended up having to stand up and say I was so sorry but could we stop filming for a few minutes so I could get myself under control again.

I think it was just seeing footage of him and me that I hadn't seen before, and suddenly being overwhelmed by a sense of his presence beside me in that moment. It was his gentle, unjudging love and his genuine pride that I suddenly remembered. All I wanted was for him to stay right there beside me on that stage.

Interviews and sound stages can be such scary, lonely places. Give me a mountain or jungle any day. In the wild there is no judgement. Ever.

On *Life Stories*, I always felt it was Shara who shone the brightest. Check it out if you have time. She was amazing. You can't hide goodness, and the pre-recorded interview she did was kind, warm, gentle and fun. I often think she would be so much better at being famous than me. The problem is that she, too, doesn't like the attention.

53

THE OTHER ISLAND

AROUND THIS TIME, Channel 4 approached our team with a concept that Shine Productions had brought to them, called *The Island*.

We all loved it from the beginning. Drop ordinary, everyday people, from all walks of life, on to a deserted island, with nothing except the clothes they stand up in and a few basic tools. Then leave them for a month and see what happens. Simple but compelling – and most definitely, at that time, original.

The original part was having no formal camera crews filming them – the cast would have to film it all themselves on little cameras. They would have zero contact with production or the outside world. It was like *Survivor* had gone hardcore, and *Lord of the Flies* had now met *The Truman Show*. It was an instant hit.

This was good for us at BGV. We now had *Running Wild* firmly established, with good solid ratings on primetime network TV in the US, and we soon had a winning primetime show on Channel 4 in the UK.

We went on to win a BAFTA for *The Island*, which was exciting for everyone involved. But I think what all of us felt most proud of was what *The Island* stood for: friendships forged in adversity and a rekindling of what really matters in life. I can sum up all that viewers, ourselves and participants gained from *The Island* in a few words: the best things in life are never things.

Every time I dropped the survivors off on their desert island, I reminded them all about the key values of a great survivor: positivity, resourcefulness, courage, determination and faith. It's the same thing I tell young Scouts to equip them for adventure and life: Please Remember Can Do Formula (PRCDF) – Positivity, Resourcefulness, Courage, Determination and Faith.

These, to me, really are the key qualities for life, and they become vitally important in a survival or adventure scenario.

We sold the format rights to the show to so many countries around the world, including France, Germany, the Netherlands, Norway, Denmark and America. Season after season the show kept smashing ratings and stereotypes.

Initially, we got a certain amount of grief in the press for season one of *The Island* being 'males only'. But the show was an experiment to find out what had happened to 'modern man'. Season two saw us introduce a female-only island. It got even higher ratings – now comparing men with women.

Then we did older people vs. youngsters; financially better-off folk vs. those with materially less wealth; and finally we added £100,000 in cash hidden around the island, with barely any rules. We called it *Treasure Island with Bear Grylls*.

Ultimately, all of these varying combinations and different seasons proved to me the same thing. That great survival isn't about gender, age, titles, wealth, nationality or upbringing, it is about character. And anyone can develop character.

What I did notice is that I have rarely met a strong person with an easy past. Often, those who had had to endure much hardship in their lives were well equipped to survive *The Island*. Because storms make us all stronger, and as they say: kites only rise against the wind. On *The Island*, all those survivors got storms and wind and rain in spades, and I have an immense amount of respect for those who endured that Panamanian archipelago.

We then floated the concept of a big primetime show with ITV called *Mission Survive*, produced by BGV. The show would involve me taking eight celebrities into the wild on a genuine adventure journey, and every two days I would send a different celebrity home. The ultimate winner would be the person I believed would make the best all-round survivor. ITV loved it.

They saw it as the perfect foil to their other survival show, *I'm a Celebrity . . . Get Me Out of Here!* A grittier, tougher, more adventurous version, and the pair of shows would give ITV dominance in the survival space on terrestrial UK TV.

We shot the first season of that show in the jungles of Central America, and it was tough. River gorges, caves, huge cliffs and limited sleep and rations. The wild always tends to sift the strong from the weak naturally, and survivalists Meg Hine and Scott Heffield did a great job on camera helping to be the eyes and ears among the celebrities for me. The show rated well. Nothing crazy, but still solid numbers, and growing. I was relieved. It had always been a goal to do a big primetime show for ITV back home in the UK, and I really believed in the values that the show promoted.

The celebrities who were still there at the end had all grown in confidence before my eyes, and it was inspiring to see a few genuine heroes emerge. On day one you would never have been able to call it. It is only when it gets tough and everyone is tired, scared, hungry and at their wits' end that you see these characters emerge. Those still standing in the final episode were all humble, hard-working, positive, resourceful team players who had learnt how to thrive in one of the most unforgiving terrains out there. And the winner was Vogue Williams, an Irish superstar, who deserved it all.

We had done our job well, I felt. We got a second season from ITV, which is always a coup, and we shot that in South Africa. Unlike *Running Wild*, where I am gentler with the guests, *Mission Survive* required me to be much harder on them. I didn't mind this. The goal was to help them grow.

Again, some heroes emerged, and despite some tense moments and tough decisions, the celebrities there at the end were examples of that never give up spirit. The winner, the former England women's football captain Alex Scott, was an inspiration to all of us – and another reminder that survival is about character over brawn.

The show also went on to earn me the title of 'Best TV Host Worldwide' at the Cannes International Format Awards, as well as 'Best Reality TV Show'. ITV were pleased. I felt really honoured to have won these accolades – I guess it's always nice to be recognized by peers within the same industry – but the real credit should be given to our production team who had worked so hard behind the scenes to make *Mission Survive* a success. Again, the unsung heroes and the giants on whose shoulders I stand.

It was slightly ironic that for both the BAFTA and the International Format Awards I actually missed being able to collect the accolades in person at the ceremonies – one, because I was away filming *Running Wild* in another jungle, and the other because it would have meant missing one of Marmaduke's school plays that I had promised I would be at.

I always get quite nervous at those big awards nights anyway, all dressed up in black tie and surrounded by so many people I don't know. That's never my calm place, so secretly I didn't really mind missing out.

The ratings had come in again and were good, but not mega, and when it came to renewing the series, the network suggested pursuing an ITV version of *Running Wild* instead. It had happened again, just as it had with *Get Out Alive* and *Survivor Games*: we start a great show, then ultimately the broadcaster comes back and says, 'What about doing something more like *Running Wild*?'

Sometimes it was a bit frustrating, but ultimately it was a compliment. After all, you only need one or two things in life to fly in order to succeed. So we adapted, and came up with the series *Bear's Mission with* . . . Our first guests were the likes of the children's author and TV star David Walliams and the world heavyweight boxing champion Anthony Joshua. As predicted, the show was cheaper and easier to make, and rated even better. So we kept making more.

The *Running Wild*-type formula kept on winning for us.

54

MASTERS OF OUR OWN FATE

WHEN WE STARTED out at BGV, running our own productions and business, I had made Del our CEO. In truth, it was too early for him to run the whole business, with all that involved, and he had the strength and humility to flag this pretty quickly. I have always loved Del's honesty and his instinct.

We hired a new person to take up the role, but it soon became apparent they just weren't right. One of the core lessons I have learnt about hiring is always to hire on character, and back people who you know will put the mission ahead of their own interests. Often, that is hard to determine just through an initial meeting, so almost invariably it means we now hire either friends, or people we know well, or a person who has come to us through a recommendation from someone we really trust.

We met C. J. Cardenas because he was helping tutor our boys while they were with me in LA. CJ had been working at the Bank of America and did some tutoring part-time on the side. He was never late, always smart, totally diligent and, above all, was clearly a good, humble, hard-working, bright guy. I asked him one day if he would ever consider leaving his corporate day job and becoming our CEO, working alongside Del. It was an intuitive but smart hire.

CJ tells me that, because I took such a leap of faith in trusting him with my future, he was determined to repay that trust with loyalty, success and dedication. He has done just that in spades.

With me based in the UK, Del and CJ were running our office in LA. The manner in which they negotiated many of our early contracts was incredibly bold, and their calm confidence in my brand set a pattern to how we do deals that remains to this day. The pair of them established ownership over shows at percentage levels that were unheard of in the media industry. They became the best of friends and made an incredible team, truly setting the foundations for everything BGV would become.

After a few years, as BGV expanded beyond just the TV shows, more and more of the business was being run out of the UK. It made sense to hire and establish a core BGV team – comms, marketing, products and partnerships – from the UK. But first up, I needed a great UK-based CEO to stand beside me and help get an ever-increasing number of fledgling projects airborne and flying.

CJ knew this was the logical next step. He was already keen to take on the challenge of running our Bear Grylls Survival Academy with Paul Gardiner, doing leadership, team building and survival training for families and businesses, so this move made sense all round. But I knew that finding a new CEO as good as him would not be easy.

All I wanted was someone trustworthy, diligent, ambitious and kind who was a natural leader but without a huge ego. Oh, and if he or she could have a legal background and deal-making spirit, that would help. Surely that wasn't too much to ask.

The irony was that in looking for the impossible I found the answer in the most obvious of places: on our doorstep, in a great friend and my fiercest table tennis opponent, Rupert Tate. Life sometimes works like that: put out good energy, expect good things, then be alert to provision.

Over one of our many table tennis battles, I pitched to Rupert whether he would ever consider throwing in his well-paid, stable, secure job for the adventure of a lifetime. He chuckled. The sort of chuckle that says: *I love the sound of that but I am a trained lawyer, with a career to pursue, and I have responsibilities.* The sort of chuckle that says: *Yes, maybe if I was twenty-five and had nothing to lose, but now? With a young family?* The sort of chuckle that says: *Damn it, Bear, my family have been institution men for generations, don't mess with my system.*

I admired Rupert so much for his willingness even to consider my offer, then to dream, and ultimately to buck convention and chuck in his safer job to join our mission. Personally, I always saw it as a no-brainer – after all, adventure, fun and friendship matter much more to me than any soul-sucking corporate ladder and big-company politics. Rupert smiled again, and rolled his eyes. 'Bear, you don't understand being institutionalized.'

Rupert taking that leap to become our CEO, and leaving his corporate career, was one of the boldest decisions I had seen – and in terms of BGV, it was arguably our smartest hire.

There was a powerful sense of destiny about the partnership, and I could feel it from day one. You don't always get so lucky in life. But I knew that in Rupert I had struck gold and was starting on a life-long adventure with him that both of us were determined would make the world a better place.

I also loved the fact that Rupert's grandfather and Shara's grand-father had been best friends; and that Rupert's wife, Henny, and Shara had been the greatest of friends since they were young girls. It all made perfect sense. I saw this simply as life coming full circle.

But above all, I had always wanted a brother to work with day to day. And in Rupert, I found one. And one of the best friends any man could wish for.

Rupert's greatest skill is caring for those we work alongside, and he leads with kindness and integrity, always setting an incredible benchmark of relentless hard work and positive action.

We are also totally aligned on putting the values of friendship and family before business – which is always refreshing for people to work with – and we always hire people on character not skill set.

We saw our job as helping empower people to find their own adventure . . . that would be the magic. And before we started to build any other businesses we wrote out clearly what we wanted to stand for. This summed it up for us both: Dream big, don't listen to the dream stealers, choose the path less trodden, embrace failure, be kind, be courageous, and never give up.

We were ready to go.

As the company grew, we brought in some other incredible, like-minded, motivated, smart people, like Lizzie Webb and Lily Taylor, who both came on board early and have been such key cornerstones of all we do. Then Mollie, Em, Ned, Adam, Caroline, Nigel, Keith, Chris, Paul, Alex and Daniel. All such strong parts of our BGV family. Plus, of course, our production and safety teams.

It is the achievement I am most proud of in my work world. How we have formed the most inspirational team, each member a huge part of what has made BGV successful. It's the collective power of great people.

We now have partners like Merlin Entertainments, with whom we built the first Bear Grylls Adventure theme park in the UK. Based permanently at the NEC in Birmingham, anyone can come and experience the best of adventure for themselves. You can skydive in our wind tunnel and scuba-dive with sharks; you can climb, do escape rooms, archery, assault courses, as well as the highest free-standing high-ropes course in Europe. There is so much going on there and it has been incredible to see the theme park launch and begin to establish itself. This sort of thing was beyond my wildest dreams when I started out on this long road, all the way back in those early *Man vs. Wild* days.

We have our Be Military Fit business that I bought with a former military buddy, Chris St George, which is helping so many former service people to start their own fitness franchises. I am so proud of that business: veteran owned and veteran run, BMF is now the biggest outdoor fitness company in Europe. BMFers train together outside in the parks and online, and it is a community of fitness enthusiasts who also love the military ethos of train hard, train with purpose, and train together.

We have also just launched our Gone Wild Festival, on the beautiful Devon coast in the UK. A family gathering where kids and parents can experience all sorts of fun adventures, as well as camping out under the stars. From wild swimming to BMF workouts, from quad bike racing and climbing to mini touch-tennis and skydiving, the aim is to make adventure fun and accessible, in a safe place for families.

A great buddy, and former Royal Marines Commando, Oli Mason, is running this with us, and he is as determined to make it succeed as anyone I have ever met. And it is all supporting the Royal Marines charity and the Scouts. I love it.

Will all of these projects work long term? I honestly don't know. But that's not always the point. The priority is always the same: set big goals that make the world a better, more fun and more united place, put killer teams in to run them, and together do all we can to help those teams deliver. It's about being unified by a zest for life and a determination to make good things happen.

Which is exactly what our education business is about. Paul Gurney, a smart, intelligent, tenacious, reformed corporate-holic, has led this for us. And it has been amazing to see it come together. It's called BecomingX, and the whole endeavour is about helping empower people worldwide with the sort of skills that school doesn't often teach, but that we all need. Everything from how to start a business to dealing with social media, debt and peer pressure; how to develop a winning attitude, how to stay fit and eat healthy, how to be a great employee and employer; how to deal with failure, risk and fear; and so much more.

It's been inspiring to see how so many of our past *Running Wild* guests have helped support BecomingX by allowing us to film intimate, heartfelt interviews with them, about their struggles and failures, demystifying what it has really taken for them to succeed in life. They, like me, know the power of real life lessons rather than just academics.

After all, the great successes in life are rarely the A-plus students at school, right?

55

FIGHT FOR WHAT YOU BELIEVE IN

SO NOW HERE we are, filming season six of *Running Wild*, now with National Geographic worldwide. I would never have thought we would do so many seasons of that show.

If you look at how competitive TV is, you will understand that 99 per cent of television pilot shows never get commissioned. Of those that do, 80 per cent fail to get recommissioned. The odds continue to stack against you, so to get six seasons of anything nowadays is rare, and a huge privilege I would never have forecast.

The fact that we now also own these shows, and produce and make them on our own terms, is a game changer. It means we can plan our shoots around other priorities, that we can film with the crew I love, where and when we want. That freedom means so much to me. It's about taking control and building a life and a career that works for us all as a team and as a family. That's a great blessing and one I never take for granted.

In the early days of *Man vs. Wild* on Discovery, their arch rival was always National Geographic. If you worked for one, you most certainly could never do anything past, present or future for the other. You had to pick your side.

Yet, as I write, I am looking at a social media feed from Discovery Channel Asia reminding me to watch *Bear Grylls Running Wild with Zac Efron* at 8 p.m., and then up flicks a tweet from National

Geographic USA reminding me to watch the premiere of *Running Wild* season five with Brie Larson. At the same time, we have the adventure I did with Kate Hudson airing in the UK. Then tomorrow we have a *Man vs. Wild* marathon running non-stop for twenty-four hours on Discovery worldwide. Maybe it's just one of those weekends, but these moments seem to happen more and more.

These are the sorts of things I could never have envisioned five or ten years ago – but then as a team we do pride ourselves on rewriting the rules. After all, many things can change with steady persistence and great people – and maybe a bit of luck thrown in as well. In the modern media landscape, this model of talent-driven creative ownership is much more common, and the industry has had to evolve. Those networks that have embraced this are winning.

If this journey has taught me anything, it is that sometimes in life you have to stand up for what you believe in, and for what you feel in your gut. Sure, it can be scary, and it requires some courage at times, but worthwhile things never come easy.

And the best things in life always require a 'never say die' attitude.

56

THE NATURAL STUDIOS

IT'S SOMETIMES INTERESTING seeing how shows grow and spread their wings. As well as how shows fail.

After *Man vs. Wild* started on Discovery Channel in America, Channel 4 in the UK also decided to buy into the format. Initially they would air the show as 'second window', as a Channel 4 show under a different title: *Born Survivor*.

Then Discovery Channel saw how the show was beginning to work so well for them internationally in all of their two hundred territories. On the back of this, Channel 4 kept investing and buying more, eventually doing multiple *Born Survivor* seasons. Networks like to back shows that they see winning elsewhere. It's a safer bet. Simple as that.

When we left Discovery and started *Running Wild* on NBC, on primetime American television, Channel 4 started *Wild Weekends with Bear Grylls* in the UK – their own similar *Running Wild* format. We had set a good pattern.

On *Wild Weekends*, we got to take some great British stars away, such as Jonathan Ross, Miranda Hart and Stephen Fry. It worked brilliantly. Then China followed, with *Absolute Wild* growing out of *Survivor Games*.

On the back of this, we started doing shows exclusively for India, then Netflix, then Amazon, ITV and National Geographic. All born from small seeds. The point is that when things work they tend to really fly, but you don't get to the good stuff without the failures.

And when it comes to failures, I have had so many. So many shows that have launched and died, never to return. But without those failures we would never have evolved and adapted and got smarter and better, and created the shows that do work. Every failure is a lesson, an opportunity to adapt and overcome.

The irony is that people tend to forget about the failures. Interviews focus on the successes; posters and billboards only go up for the popular stuff. But the truth is that those 'successes' are really only the tip of an iceberg of failed projects. And if I am proud of anything, it is that we kept going through those failures.

Our boys were asking about this the other day, and I actually wrote out the total number of different episodes and seasons of shows we had done. We counted thirty original formats over forty-five seasons of shows, and over 450 episodes filmed. Yet the shows on that list that survived beyond one or two seasons and that are still going I could count on one hand.

The point is that the hit rate of successes to failures is overwhelmingly low. In other words, you have to get used to failing, get used to hurting, and treat them both as a solid sign of positive progress.

The same applies to the level of fees I used to get paid. For the first ever TV show we did, for example, *Escape to the Legion* for Channel 4, I was paid a few grand for a month-long, 24/7 shoot. Not that I was complaining. But I do remember flying out in the back row of the plane, looking at the execs sat in row 1A up front and thinking: *Hold on. I came up with this idea, and I'm the guy who's going to be hosting it and going through everything with these recruits. How come I'm back here and they're up there?*

There is an element of the fact that for many of those early years I was simply doing my time, and learning my trade, but there is also this truth: you get what you negotiate in life. No one is going to hand anything to you on a platter, ever, and in the world of TV, producers exploit hosts, 'talent', actors, whatever you call them, all day long.

A month of being thrashed around the North African desert by French Foreign Legionnaires to show what it is like to go through basic training with France's band of mercenary soldiers was a baptism by fire into the world of filming and TV. It was a tough, grim experience, but I made a few lifelong buddies from some of the recruits and crew, like Bobby Abedeen and Will Collis, both of them still great friends; and camera operator Paul Mungeam (Mungo), a legend of a man as you've already seen, and still our lead cameraman to this day.

That first show, *Escape to the Legion*, was our first hit on UK terrestrial TV, and in many ways that show launched my TV career. It opened the door to allow me to do a big US show for Discovery Channel. And therein was born *Man vs. Wild*.

I look at it and can clearly say I got pretty lucky. We hit two solid bullseyes with those first two shows, and that momentum carried me through the many failures that followed. Then again, as they say: luck is a dividend of sweat.

The fees slowly grew over the years and seasons, but even when *Man vs. Wild* was at its height I still had zero ownership over the shows and was getting paid relatively little for the risk, danger and time investment required for each episode.

I think my fees started at around £15,000 an episode, which back then took around ten days to shoot. Great money in normal life but in the world of TV I knew that the big bucks were definitely not being earned by those taking the risks and doing the dirty work. The 'head sheds', execs and networks would be pulling in the millions, and a tiny per cent of that was trickling down to those of us with our faces in the dust.

As I've said, I was determined to change that. Not just for me, but for our team, for Del, for our crew, and for my family. I was determined to change it for the time when ratings fell, or my legs and back started to give way. I was determined not to be left with nothing, thrown on the pile of ex-TV hosts who were just over broke from their Just Over Broke (JOB). That pile, I was learning, was a big one.

Most hosts or actors get wise when it is too late. You have to get smart early, when you have leverage and power, if you are going to fight for what is right. Otherwise, you get killed. We did. We took on those battles. They were worth fighting.

From those beginnings, Del and I eventually grew the brand and fees to the point where I was earning almost half a million dollars for an afternoon's filming. As Del noted, 'That makes you better paid than Daniel Craig as 007, and almost every other film star out there. Good job.'

Together, we had simply built a stable of adventure shows that delivered consistent, strong international ratings for networks around the world. But still, to command that level of fee was uncharted waters, for both us and the networks – not to mention that we also co-owned our rights and formats.

It was always our hope that some big player, whether Amazon, Netflix or another big global production house, would look at acquiring what we had built. The vision Del and I had was simple: imagine if we kept doing all this but with a bunch of other adventure talent and hosts as well? We could build the biggest home of adventure production in the world . . .

ACF, a media investment banking firm, helped us put this vision together in terms of what the company could potentially look like. It was based on what we were currently doing, but expanded to bring in other adventurers, and building shows around them and their aspirations.

The journey to find and choose a buyer was an exciting one, and we had many turns where we said no to some big players. I was just determined not to be dazzled by offers only then to get tangled up in a bad contract. My fear was being controlled and owned by some other party. I had been there before.

In many ways, this meant rewriting the way TV and media deals had historically been put together – but challenging the status quo was now firmly in our DNA. I have Del and Rupert to thank for supporting this way of thinking. Warriors beside me – and without them we would never have built our first studios.

But eventually, after months of negotiating, we launched and then sold half of what we called The Natural Studios, to the Banijay Group, who in turn have proved inspired partners for us. Always encouraging, always empowering.

We sold the business for an upfront sum that was beyond anything I could ever have hoped for when we started out. Especially when I think back to those days of dragging myself through swamp after swamp. But sometimes it's important not to look too far into the future. The summits can be daunting. Instead, I always knew I had to build a great team, walk the path less trodden, and never give up. Those three factors are everything.

The memorandum for the deal we did was appropriately called Project NGU. It was living proof that life often rewards the dogged and the determined, more than just the skilled or the talented.

It's worth bearing in mind that historically, as one kind researcher wrote to tell me once, my family comes from a long line of impoverished Catalonian boat builders in Spain. The researcher explained that Grylls was a Catalan name derivative, and that some of my ancestors had come over with the Spanish Armada naval fleet, got soundly defeated by the Brits, then shipwrecked and eventually washed ashore on the UK's Cornish coast.

With our Grylls dark skin and big noses, Cornwall was where my ancestors settled, married and had children.

We might have been seafaring and happy, but the truth is that no one in our family had ever made much money. Not that money means everything, it doesn't. But I still felt a quiet sense of pride that my modern-day family had somehow just got paid extraordinarily well for the privilege of having some great adventures with great friends. For a long time after we signed the papers, that truth felt surreal to me.

I've learnt in life that to make any sort of money is hard work, and invariably it involves a lot of luck – but it is also a privilege. One that you have to try and use wisely.

As someone once said: 'Money is like manure – it only works when it is spread around.'

I always loved that.

57

20/20 HINDSIGHT

AS I WRITE this next chapter, I am back in Los Angeles. I am here to see the chairman of MGM, Mark Burnett, about our Amazon Prime show *Eco Challenge: World's Toughest Race*. I am due to review some of the edits, and also to deliver a keynote talk to a big US software firm.

I don't come here nearly as much as I used to, but The Natural Studios and Del still work from here. He often jokes that he gets way more done when I am away. 'No disrespect,' he always adds with a smile. I don't doubt that he is right.

I find it strange coming back to LA these days. I can look back and see my early career here for the time that it was, without the overwhelming fear of it all going wrong around me – which was a very real concern that I had every day during that time.

I remember so clearly the smells of the city, the sounds, the design of the sidewalks, all from those early days with Dave Segel and Del, racing from meeting to meeting, between lawyers and show pitches. Always hustling. Almost always on the back foot. But trying to learn, to listen. Trying to share a vision of how adventure can inspire people.

It was a whirlwind every day, and at the time I don't think I ever stopped to think for too long. I was jostling, always busy. *Keep the faith, BG. Stay strong, get up early, be on form . . . do your best.*

One time, Del and I were racing across town to meet Simon Andreae, the then head of Discovery North America, for what he was calling our last chance to resolve our issues 'amicably'. Once past this point, he told me, there would be nothing more he could do to help. If we still insisted on ending *Man vs. Wild*, the lawyers would take it from there.

I was nervous and under pressure. And losing.

Del's battered 1987 BMW was a perfect reflection of how I felt. It was overheating and losing power, fast. He tried to coax it along the highway and we just made it off our turning before it rolled to an ominous stop. We still had 4 miles to go.

The only way we could get the car going again was to let the engine cool down for five minutes; then it would restart and limp on for a few hundred yards, then stop again. Eventually, though, it wouldn't even start. We needed some other way to cool it down. And fast.

I grabbed an empty Coke bottle from Del's backseat and before he had asked what I was doing I leapt over the wall of some Beverly Hills mansion, landing quietly in their garden. I found the outdoor tap, filled the bottle, then leapt back over the fence on to the sidewalk.

'Let's hope this works,' I said to Del as we refilled the leaking water tank. 'We need to be at this meeting.'

The car started, and we limped the last couple of miles into the Beverly Hills Discovery offices.

The meeting was amicable but we didn't resolve anything. Instead, we held firm and refused to buckle and restart *Man vs. Wild* on their terms. This saga was going the whole way. But as you now know, we got there in the end – mainly with a heady mix of luck and tenacity, plus a little diplomacy at the right time.

Back then, it was a frightening place to be. I vividly remember that feeling. Hustling. Fighting. Scrapping. They say that fortune always favours the brave – it rolls off the tongue so easily. But being brave when lots was at stake, including my future and my family's, was hard. Somehow we kept going.

While Del and I were racing around, the family were back at the rental home we had near the beach in Malibu. It was a beautiful respite from it all, every time I got home. I would often sit up late and watch the ocean. The waves. Always measured. Always strong.

Hold it together, Bear.

I look back, though, and realize that I never really let myself face the true emotion of that time, that if I had stopped for a second and thought about it, the dire situation of losing everything might have overwhelmed me. I had long ago burnt every bridge in terms of having a regular career, and now I had just walked away from the biggest gig anyone could ask for in TV.

I had a lot on the line.

But facing up to that fear was something I didn't even dare acknowledge. I was too scared that if I started to think about it all, I would crumble. I told myself to have faith, to keep trusting. I had this deep-rooted belief that somehow good would prevail. So I kept my head down and just kept going – and believing.

But the fear was always there.

Fear about whether I was crazy to have tipped over that apple cart, to try and go solo and build our own production studios. Fear about whether I would be able to look after my young family.

Fear. Fear. Fear.

But I refused to let it in.

It was as if, during those months, I had been given a mantle of courage that was totally unjustified. As Del told me recently: 'BG, you carried us all through those storms. If you had not had faith, we would have been screwed.'

Now, sitting on the other side of the table – coming back to LA, having established our production studios, where we own our shows, having built a brand that stands for so much good – things are so different. Yet in some ways still the same.

I come off stage after my talk and I say my goodbyes to both the CEO and the security team that have been escorting me. The security part never feels necessary but the speaker agency and the software company insisted I have it. I stopped arguing that sort of thing a while ago. Personally, I think it just makes everyone feel more important – that their guest speaker is some big shot. Even though I am not.

I thank them all, again, then walk a few blocks to get a coffee, on my own. Some quiet space again. I always crave it. Somehow even more so in the cities.

But it is only then, when I'm walking down the street, that the emotion hits me like a tidal wave.

I know that I am suddenly feeling what I should have felt back then, when I was in that hole all those years earlier. Fear. I don't suppress it any longer. As I walk down the street, I let it wash over me like a balm.

Feel it. Embrace it. It's OK now.

No longer do I have anything to be scared of.

We prevailed. Our team is strong. We did it.

This is part of why I am so loyal to Del: he was with me in the tough times, when we were the little guys. Up against it. The David to the Goliath. But he helped steady my aim, and found me the perfect pebble for my sling. And now we can look back on those times with pride.

When I remind Del of all this later that evening, I see tears in his eyes. Ours is a friendship forged through the storm. And it feels good to remember how far we have travelled.

My promise is never to stop being grateful. And always help the hustlers on their way up.

Because we've been there.

58

GETTING YOUR KICKS

HAVING FUN IS important. After all, if you're lucky enough to get a break, and against the odds you experience some wins, then you're a fool if you don't also sometimes stop and enjoy the moment.

Enjoying the moment means different things to different people, but for me I have never been a flash car or designer clothes sort of guy. It's just not what excites me. (Although, don't get me wrong, I am forever grateful to Land Rover for their sponsorship over many expeditions and years.) It is just that I get my kicks in different ways.

Skiing has been a great love of mine for many years now. It's that mountain spirit that has always been a powerful part of my life. Mountains demand respect and humility, if you're going to get along. I like that.

The small village where we spend part of the year, in Switzerland, is a unique mountain community of talented, family-minded, former mountain or ski guides and adventurers. Both Huckleberry and Marmaduke have grown up going to school out there for some of the winter terms, and they love it. It has allowed us to build a great community of adventure-loving folk, including BASE jumpers, freeride skiers and top climbers.

Shara and I did our ski instructor qualifications together a few years back, yet we both still feel like rookies compared to so many of those we share the mountain with. Whenever I am asked if I ski well, the answer is always 'I'm enthusiastic', because compared to many – and nowadays I include our three boys in this category – when it comes to skill level, I am a 'survivor' not an expert.

The first time I watched Jesse doing a perfect layout backflip on skis, flying over the top of me, from a steep cliff cornice, I knew my days of leading the pack had gone. The other two boys, likewise. But the mountains and fun ski days together have been a wonderful thing in Shara's and my life as a family.

I also like the challenge, the never-ending learning curve, and I hope the days of 'always that little extra' will never end.

Only this year we had some of the world's best big-cliff freeride skiers with us, and Jamesa Hampton, one of the up-and-coming young Kiwi skiers, was leading us down a pretty committing chute on some cliff face. He hit this lip and pulled a beautiful somersault, landing nimbly. He called up to me to say all was good and the chute clear.

I hit it hard, but my tips caught an ice lip, and before I knew it I was catapulted forward through the air, leaving my skis far behind me. It was steep terrain and luckily I landed in deep powder, and all was fine. Later that evening, Jamesa and everyone were chuckling about it. He said, 'I've seen a lot of big tumbles before, but I will never forget looking up this chute and seeing you, Bear, flying through the air, in Superman position, with no skis on.'

We all laughed.

'But, respect. Total commitment.'

That for me is life and the mountains. Give it your all, fall down often, get back up and laugh at yourself, with good friends – and, of course, never give up.

And remember: *You're not eighteen now. Be smart.*

The fact that Jesse and I can also paraglide and speed-fly on mini parachute canopies off some of the most spectacular peaks on Earth, right from our doorstep, is also a big draw to the Alps for us. Jesse now flies solo alongside me – not to mention that he can skydive, paramotor and BASE jump as well. (Cool skills to nail at sixteen.)

Those moments, skiing off some truly epic mountain faces, then soaring through the air at high speed on tiny speed-flying canopies with Jesse hot on my tail are special.

Likewise, the calmer flying we do in the summertime as a family. Those sunset paragliding flights in the Alps with all five of us as a family, airborne simultaneously, riding the end-of-day thermals, laughing and spiralling down alongside spectacular ravines and over forests.

We fly with two great local buddies, Mike and Stu Belbas, who pilot Shara, Marmaduke and Huckleberry in tandem. That way we can all be airborne together.

Great times, great friends, great moments. I live for them.

Then there are the times when we are back in the UK – whether up on the island or on our London houseboat – and you might wonder where I get my kicks then. Well, no doubt the Be Military Fit (BMF) training I do most early mornings, often with Rupert before he settles into his BGV day, is an anchor in my life. Short, hard sessions, often also done online, as part of our BMF community of fitness enthusiasts. I love them.

Likewise, the endless mini touch-tennis battles that I have with Rupert, where we hustle, sweat and pour our hearts out on this little improvised court. Moments of effort and escapism, where the softer ball and smaller court allow us to play at a level where we appear as intense as any tennis pros.

At times we have been known to cover over 5km of shuttle sprints back and forth in just one thirty-minute touch-tennis session. And it all plays its part in building speed, cardio – and, above all, friendships. The great currency of life.

Meanwhile, Marmaduke, aged fourteen at the time of writing, is currently determined to join the Royal Navy, and he is as original and gorgeous a young man as you will ever meet. He is as fun and inventive as he is kind and authentic. One in a million, for all the right reasons. And none of us can handle our small ex-RNLI inflatable, which we use to cross over to our island, like he can. A true rough-water expert. Especially if there is the promise of a full English fry-up in the beach hut on the mainland with friends at the end of the crossing.

And as for Huckleberry, the youngest of our three, he's sport obsessed, with the abs and the talent to pull it off. Yet so cosy and sensitive, as only children can be. He out-skis us all, and is close behind me now on the touch-tennis. The life of a parent, eh? You teach, love, grow and encourage, then get bettered at everything in return for the privilege.

I love it.

59

STAYING WINTERED

AMID ALL THE fun that we have as a family, there is one truth that I always know we must adhere to: *stay wintered*.

It's the concept of keeping the inner part of you tough. And toughness is like a muscle: the more you work it, the more you visit those difficult moments and pain-filled endeavours, the more resilient you become to the hardships. And in the game of life, resilience is king.

Last week I flew back from Switzerland to the UK in between filming. I had wrapped up a winter shoot in the high Alps and was heading on to the plains and jungle heat of southern India. In between I had a day in the UK to go and see Jesse at school, drop and collect some gear from home, do an online talk for American Express, smash a workout and get back to the airport.

By the time I got to where I was spending the night, before heading back to Heathrow at dawn, it was past midnight. Being mid-February, there was a frost on the ground.

There is a lake nearby that I know well. I knew it would be amazing under this cold, midnight sky. I dumped all my bags, put the kettle on, then stripped down and ran the 200 metres down to the lake.

It started to rain.

I dropped my shorts, took a deep breath and dived into the blackness.

Ice water is always bracing, but also intoxicating. Especially at night, alone.

I sat in the shallows, with just my head exposed. The air felt warm, compared to the freezing water. I breathed deeply. Slowing everything down. In my head as much as in my body.

Let the cold heal.

And never forget what brought you here.

Stay wintered. Keep that inner resilience muscle tested and strong.

I like this power of nature to shock, heal, challenge and restore. And few things do that like an ice-cold lake on a winter's night.

I have had so many of those moments for 'work' – whether on *Man vs. Wild*, escaping frozen lakes; or crossing icy rivers in the depths of winter with the pop star Nick Jonas, or the actor Channing Tatum, on *Running Wild*. So many. There's always that initial sense of trepidation from those with me, followed by exhilaration when we make it out the other side. Again, the healing power of nature to bring confidence and empowerment. My happy place.

This concept of staying wintered is something I have spoken to our boys about since they were small. Not running from the failures, nor solely seeking the secure, the comfortable. It is about searching out the hard stuff. It's about knowing yourself. Becoming familiar with fear and pain. Knowing how to find a way through them. These are the things that empower young people for life – not just good grades, effortlessly won.

School doesn't always teach us about life. The raw reality of how tough it is. How it will kick the hell out of you many times, and force you to face many struggles on your own.

There are few safety nets in the big world. We have to equip our children to deal with that truth. Teach them the value of failing, of trying, and of not succeeding first time. Even second, third, fourth time. Get used to aiming high and getting knocked down hard. Teach them the one skill that really counts: resilience.

How to get back on your feet, once more.

This, to me, is the very real danger of being a 'success' at school. Being too good, too easily, means there is no struggle. But it is the struggle that always makes us strong.

This is the stuff we teach at our Survival Academy and also with our BecomingX Foundation. It is a message I am so proud to champion. It is why, despite any success, I will always do all I can to stay wintered. Maybe not as much as when I was twenty-one – even thirty-one – but still wintered.

And I inspire our boys to seek it out, too. To know the power of doing difficult things. Keeping in touch with that inner steel. Train that 'effort' muscle.

And never pass a chance to swim in ice-cold water.

Preferably in the dark.

60

SPEAKING TRUTH

THE ART OF communicating well is never truly mastered – we constantly learn, and can always improve. The many thousands of talks I have given over the years, including a few disastrous ones, are where I did my time and learnt to tighten my message. To communicate better.

The key is always speaking from a place of true vulnerability. If you can get that part right, then the rest is detail.

Recently, I was speaking at a convention alongside President Obama. I hadn't seen him since he had finished in office. It was fun to see the whole Secret Service machine back in action again. I had forgotten how crazy tight security is wherever he goes – even though he was no longer Commander in Chief.

Backstage, in a room set aside for us both, we got to hang out a bit and chat. He was the same warm person as before. For me it was especially fun to be able to introduce him to Rupert and Lily who were travelling with me. Life is all about sharing cool moments with those we love.

I had delivered my talk an hour or so earlier, and it was now POTUS44's turn on stage.

It felt so strange to be sharing the experience of presenting at one of the big conventions with Obama, and I kept thinking how far my own journey, in the world of corporate motivational speaking, had come.

Again, it had grown from simple, stumbling beginnings.

In fact, one of the first talks I did after climbing Everest was to a hall full of locals down in the Isle of Wight, where I grew up. The slide machine was perched on a stack of books and I had a bundle of notes in my hand. I was so nervous. I mumbled my way through it, dropped my folder twice, and the slide machine, at one point, seized.

At the end, a well-meaning elderly gentleman came up and said that I would most definitely benefit from some training on how to deliver a presentation. He knew someone who could help if I wanted it. It was the ultimate put-down. Still, I took him up on the offer. I couldn't get much worse.

With practice, I slowly learnt to get better. Less nervous. More concise. Clearer. Slower.

Breathe, Bear.

Over time, as I started getting ever bigger jobs and being better paid, I remember feeling a pressure to dress smarter and to make my message more corporate and formal. It took me a while to realize how wrong that was. I wasn't being hired to be just another suit. They had enough of those in the audience. I was there to tell an honest story and share what had helped me at difficult times.

In truth, it's never about what we look like – regardless of what they tell you – it's always about your heart. It's fine to wear a T-shirt and jeans if what you are saying is powerful, intimate and straight from your soul. As the founder of Scouting, Lord Baden-Powell, once said: 'It's the spirit within, not the veneer without, that matters.'

But to embrace that takes confidence – and it took me a while to get there.

The best advice I ever took on board wasn't from any public-speaking training company, it was from Sir John Mills, the legendary English actor. Just a few years before he died, we happened to be speaking at an event together. He was on first and I was to follow. Backstage, I asked him nervously what a lifetime of communicating to so many people had taught him.

He summed it up thus: *be sincere; be brief; be seated.*

No wonder he won hearts all over the world – and I've never forgotten those words. They're so true. You don't have to be formal, or even funny. In fact, the whole notion of starting with a joke or anecdote is so flawed. It's a smokescreen. And when you hide, you lose your power.

Just be sincere, be brief . . . and be seated.

Over time, I have finally developed the confidence to speak from the heart. To share the struggles, and the doubts and fears. True vulnerability is painful to share, and it takes confidence, but it is the heart of all good communication. It's as simple – and as difficult – as that.

Nowadays, I try to speak about fear and failure, about friendship and kindness – not always the sort of words you hear much about in corporate settings, but they are words that matter – whether I'm addressing a company or a school or a government.

I often get asked what it is like speaking on a stage to sometimes as many as twenty thousand people. And, of course, it's scary, and hard, but I now realize that when it comes to delivering talks in front of large audiences, if it's easy then you're doing something wrong.

Now, when I am stood waiting backstage, I will often sneak a peek out from the wings at the audience. It will always be a frightening sight – that's just how it is – but I know my message, and I have faith in its ability to encourage, touch and hopefully motivate people. I have seen it so often.

In the early days of speaking, I always felt a need to tell stories that built me up, as if somehow I had to establish the credibility to be there. I guess because no one knew who the hell this young guy was, coming on to the stage in ill-fitting clothes, with a badly knotted tie and hair scrubbed down awkwardly across his head.

Left: A special moment of connection with Prime Minister Modi of India – wrapping him in a poncho to try and keep him dry, midway on our journey together. The wild is always the ultimate leveller, and I love that.

Below left: At this moment the Indian Secret Service team was going a little nuts, running along the river bank, worried their PM was about to drown. I had promised them that the raft wouldn't sink, but at this point I wasn't so sure.

Bottom: A shot taken after the Prime Minister Modi adventure, including many of our incredible crew: Pete Lee top left, Stani Groeneweg in front of him, Jimmy Goddard under the boom, Delbert Shoopman red hat, Ben Kenobi behind him, Rob Buchta next to me, Dan Bowring to his left, Paul Mungeam and James Blythe kneeling front right, and the late Josh Valentine far right.

Above: Probably my favourite work photo. The wild always keeps things real and encourages us to laugh at the simplest of things. In this case, the Chinese stars and myself were about to eat some massive bull's testicles.

Right: This photo sums up the dynamics involved with taking multiple celebrities on a journey. You've always got to be on your game. This was taken deep down some gorge in the Chinese jungle.

Left: My time as a Trooper with E Squadron 21 SAS: these were some of the proudest days of my life. Those few years gave me not just skills, but, maybe more importantly, a quiet confidence that I didn't have before.

Below left: I feel so proud to represent the Royal Marines Commandos as their Honorary Colonel and to champion all they do around the world. Heroes one and all.

Below right: Dressed in full ceremonial 'Blues', ready to take a passing out parade down at Commando Training Centre, Lympstone. I always feel my father's spirit with me when I'm down there. As a former 'bootneck', he would have loved to have seen this.

Left: Yomping alongside some Royal Marine recruits, closing in on the end of their final 30-miler Commando test.

Above: My happy place – flanked by unsung Scouting heroes.

Below left: Alongside HRH The Duchess of Cambridge at Windsor Castle, for the annual Queen's Scout ceremony. This was Catherine's first official solo engagement and such a special day for the many Scouts present.

Below right: Having just won the legendary King's Cup and been passed the trophy by Prince William, I then so nearly dropped it. It was close to being a very expensive moment.

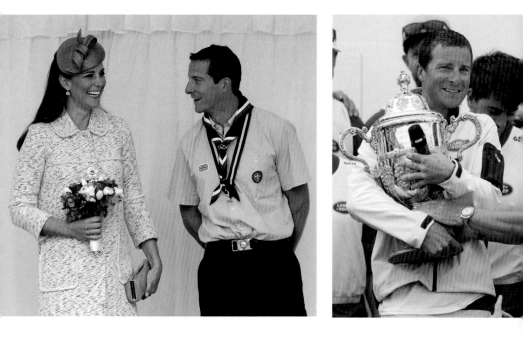

Top left: Shara and the gang.

Top right: Shara and I have always made a good team. It's one of the key adventures to get right.

Above: A family day out in our refurbished boat.

Left: The island gives us that simple time where we can all just be together.

Above and below: Truly the place I love more than anywhere else on earth. St Tudwal's Island West, North Wales. Off grid, offshore, and the best thing we ever bought as a family.

Below: The boys have learnt the value of each other and adventure on the island . . . as well as of waterproof clothing.

Left: Shara on the island. No words necessary.

Below left: Simple moments – always the best.

Below right: With Marmaduke.

Left: Marmaduke, Shara and Huckleberry in the Alps, where we spend much of the winter. Huckles with a busted nose after a backflip on skis went wrong. Now his nose bends left like mine.

Below: So many favourites in this shot, I don't know where to begin. Island, Shara, dungarees.

Above: Photo shoots are never a happy place for me, but our team have learnt how to corral me into them once a year under sufferance. This front cover of *GQ* (*left*) was an honour to do, but, still, we shot the whole six pages in under twenty minutes. Job done.

Below: Life motto: *Always look up, smile when times are tough, know that love wins, and, above all, never give up.*

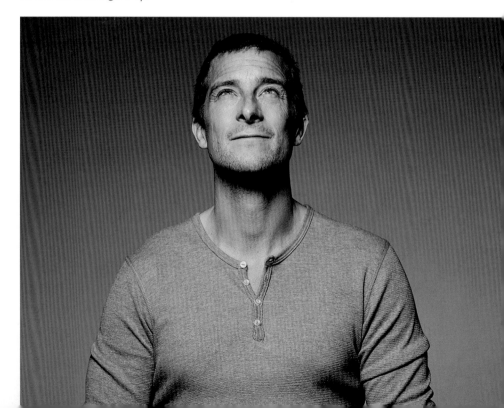

But I no longer have to establish any authority, consciously or subconsciously, to be on that stage. Instead, TV has naturally built that for me, before I ever get to any podium. I can feel it. But with that comes expectation.

People expect Bear Grylls to be a certain way.

They expect adventure and heroism and unrelenting positivity.

But that isn't how real life is. Certainly not mine. The truth of my journey is rooted in struggle – but it has been that struggle that has made me strong.

That's the story worth telling.

If anything, I find I have to unravel and almost deconstruct that adventure-guy image. That's not to play everything down with false modesty or understated Britishness, it's about having the courage to show that I'm not this all-action hero figure who can conquer anything. To share those struggles and fears which, in truth, lie at the heart of all great adventures.

I mean, what is adventure? It certainly isn't about easily overcoming an obstacle. It's the ability to keep going, to try and stay calm against the odds, to know those moments of raw terror yet still be able to edge onwards, albeit nervously, into the storm. Adventure is about moving towards the difficult, despite the tide of chaos and change that often threatens to overwhelm us all. And doing it with positive purpose.

The Royal Marines Commandos call it 'cheerfulness in adversity'. The Scouts term it 'courage in the face of adversity'. It's about being able to look disaster in the face and finding a way to prevail.

That's true adventure.

And it comes in many different forms, across many different arenas: our work, our relationships, our internal struggles, our dreams and ambitions.

That's the message I want to communicate.

61

FEAR IS YOUR FRIEND

MAYBE THE BEST way of doing this is simply to take you backstage with me, here and now. Live, so to speak. Standing beside me . . .

You ready?

The music is a thumping bass.

The sound of the audience shuffling around, and all the chatter, mingles and drifts backstage in a wave of energy.

Audiences of over ten thousand can't fit in most normal theatres or conference centres, but across America and Asia there are a handful of mega venues that transform into giant forums where some of the world's biggest companies and gatherings of industry get together.

You know when you are at one because no one goes anywhere backstage without a golf cart. The distances that crew have to cover as they move around at these huge conventions are vast. It's pretty amazing to see. Like another world.

Endless long tunnels, endless tea urns, green rooms, sound engineers, and everyone always dressed in black – earpieces and radios in, always busy. Cables everywhere. Personal torches flashing as crew move around. Hushed tones. Clipboards.

Black flooring with white tape and arrows marking the way to various entrances to the main stage. Huge amps, speakers, distribution boxes 8ft high, flashing insanely with LEDs. Rows of technicians on headsets behind monitors.

Then, once you are through all of that, there is the no-go zone, close to the main stage. Everything is slower and calmer here. This is where you wait to go on. It's calm because all the hard work is done by this point.

It's the final area before the actual stage itself. This is the only small space in the whole arena – 6ft by 6ft of black floor. That is your world, while you wait for the stagehand to give you your final cue.

I am mic'ed, and shuffling with nervous tension. But I'm as ready as I can be.

A finger goes up. The stagehand is listening to his headset for the cue. The noise of the music and crowd is almost overwhelming.

Five, four, three, two, one – GO.

A smile from that final crew member as he slides the screens open. There's a blinding flash as you first glimpse the stage lights, blinking after the darkness of backstage, and then you're on . . .

It's time to deliver.

62

FAIL FIRST

WHAT FOLLOWS IS a word-for-word transcript of a talk I gave recently. Much of this is my story before TV, covered in my 2011 autobiography *Mud, Sweat and Tears*. I wanted to include this talk to show how these foundations set in motion so much of my journey since.

And if you're unlikely ever to hear me live, well here you go.

It's show time.

Thank you. That's a very kind introduction – really – and I guess it's always nice to have the best side of us put forward . . . but there's also a little part of me that thinks that maybe a more real version of an introduction might actually go something more like this:

'Bear Grylls will be with us momentarily – it seems that he's currently gone walkabout . . . again. Bear was last spotted backstage nervously scribbling notes for this talk . . . He has since visited the bathroom four times.

'Bear Grylls watches far too many YouTube clips of epic ski fails – actually pretty good – and sometimes he catches glimpses of himself on TV and wonders how on earth all this ever came about.

'Bear also wonders if people know quite how much he struggles with the recognition side of things . . . and that he much prefers hanging out with good friends, being treated like a normal guy, than going to any fancy party.

'Bear also drinks thirty cups of milky tea every day and remains convinced that this is entirely appropriate for an adventurer.'

All very true.

But, I guess what I really mean is this:

That the man standing in front of you is often full of nerves, and I know struggle, and I've experienced many moments of self-doubt, but I look in that cold mirror and I see four clear things that have made such a huge difference, four words that have come to define so much of my journey.

These things have been painful at times . . . but they've forged me. Quietly. Over time.

And by the end of this session together I hope you'll understand why.

So first, and arguably the most important one: the Failures.

And this is where I want to start: stood in the rain, at the foot of a windswept mountain in the Brecon Beacons. Soaked, shivering, and sick with nerves about what lay ahead over the next eleven months.

This was day one of British Special Forces Selection to join 21 SAS.

We were a band of ninety recruits – or 'volunteers' as we were called – and if I'm honest, I had never felt so out of place among these huge, muscle-bound soldiers. I had that dreadful feeling that I'd volunteered for something way beyond my capabilities.

The first time I tried for the SAS, after six months of feeling beaten up, I was failed. 'Returned to Unit', as they call it. The raw truth was that I wasn't fast enough, smart enough or good enough. And that truth hurt.

People often don't know that I failed SAS Selection first time. We all like to gloss over our failures. But my failures far outweigh the successes, many times over.

The failed expeditions, the failed projects, the near-death moments that still come to me in the night sometimes. But those failures have made me, and built resilience. They've forced me to adapt, to get stronger. Inside and out. Because there's no shortcut to your goals that avoids failure. Failure is a doorway you have to go through to succeed.

And as a young soldier, I was determined to retry for the SAS. To go through the notorious Selection process again.

Such a big part of me dreaded the pain, the blisters, the unknown, the fatigue. We all seek comfort. Part of being human. But comfort isn't comfort if it isn't earned.

Soon, another ninety volunteer recruits were lined up. This was now the winter Selection course, reportedly tougher still. And then the hard work began. Again and again . . . night after night . . . dragging ourselves and our backpacks across the mountains.

By the same stage that I'd failed first time, we were down from ninety to eighteen recruits. And I'll always remember one particular Welsh mountain peak, the last point on a long route march that had been going on for ever.

All of us were wrecked, feet swollen, backs arched under the weight of the packs. And as we descended from the summit we could see the faint glimmer of the lights from the Army trucks at the foot of the mountain.

Those lights meant rest, and another of the many phases complete.

But not that night. Because when we finally reached the bottom and approached the trucks, the engines burst into life and they drove off.

The steam of sweat was pouring off us all. We stood slumped over our rifles, watching as the vehicles disappeared into the black. We were done. Like the walking dead.

That comfort voice was in overdrive. 'I told you this was a really dumb idea . . .'

The Corporals then emerged to tell us that the trucks would meet us the other side of the peak. Back over . . . again. How many times had that now been?

Then the silence was broken, and one of the recruits dropped his pack and slumped to the ground. Then another. And another. And at 3 a.m. that morning, in the bleak rain, six recruits in total gave up.

But do you know what happened just after that?

Before we had barely gone 200 metres back towards that mountain, the trucks returned to collect us. The Corporals had simply wanted to see how we'd react when truly empty. Because Special Forces Selection reduces people, and looks at how you deal with that pressure.

This time, by the end, all those months later, only four of us remained. By then, those men beside me were like brothers. To look at, all pretty ordinary. Regular in stature, yet mighty in spirit. Still best friends to this day.

But you know the real irony? That out of the four of us still standing, three of us had failed at our first attempt, but then got it second time. The first failure was actually the key.

So embrace those failures. Don't run from them.

Know their role. Essential markers, doorways we have to pass through to reach our dreams.

So that's the first thing. The failures.

Next – and this one shapes us all in some way – is Fear.

63

FACING FEAR

Life is scary sometimes. And all of us face battles. Battles of con-fidence, and nerves. Battles where we have to face some giants. Sometimes big ones.

And those giants can be really frightening.

But it's a universal truth that whoever we are, life will test us physically, mentally, emotionally. How we react to that testing determines everything. Because life doesn't always reward the brilliant, or the talented. It doesn't really care about the exam results, the good looks, the letters after a name. No, life rewards the dogged, the determined, those who can walk towards their fears despite the giants.

After three years in the military, I had a moment that changed my life for ever. It was a routine parachute jump in Africa.

It's getting dark, and I'm free-falling past 4,000ft. Time to pull.

I look up, to check the canopy is open, but something's very wrong. I'm struggling to control it. I'm still falling . . . way too fast.

I'm too low now to use the reserve. I brace for the impact. My world goes black.

I remember the African hospital. I'm alive but I can't move. Then the doctors, the pain. My back is shattered in three differ-ent places.

I remember the blurred journey home. Then the months in rehabilitation. The night terrors, always about falling, always out of control.

You might wonder if I've ever jumped again? The answer is all the time. Part of my life.

A choice. To embrace the difficult.

Don't get me wrong, I dread the jumps, and I dread the nights before. That roar of the wind as the plane door opens. Terrible! But I know the answer to fear: Face it. Don't run. Use it. Become friends with it.

Because life and experience have repeatedly shown me that when we edge towards our fears, they so often melt away. That the only true way over our fears is right through the middle.

The wild says it too: you've got to face the wolves, swim strong if you're surrounded by sharks. Because when we run, everything gets worse. The exception to the rule being a saltwater crocodile . . . don't go towards that one. Especially not in murky water!

Anyway . . .

You might not also know this, but I actually really struggle with rooms full of strangers. I know, the irony isn't lost on me, trust me. But I think it's because I know I'm not as strong as people might expect . . .

But that's OK. We all have our stuff, and our fears are just part of life – part of what makes us real, and relatable, I hope. You know what they say: light can only shine through cracked vessels . . . because those cracks are really the story. Like wrinkles.

My wife Shara and I have three young boys, and I say this to them when they feel the crow's feet around my eyes. They're nothing to hide, each one a story, an adventure . . .

This one? Trying to cross the Arctic Ocean, five of us in a small, open rigid inflatable boat.

It's 3 a.m. and we're in a force 9 gale, 500 miles offshore. There's ice all around. It's pitch black. We have huge waves and white water crashing on us like houses. Twice our small boat teeters on the edge of capsize.

And the storm's getting worse.

I see my best friend looking at me with those eyes: We go over in this, we die.

The big moments in all our lives leave marks. And if I'm honest, I'm pretty covered in marks. But life doesn't require us to be perfect. Perfect doesn't exist.

Life just says: Keep moving. Always forward. Use the fear as a driver. An emotion to sharpen us.

And as for the marks, the scars, the wrinkles . . . they make us who we are.

64

FINDING THE FIRE

So, we have the failures, the fears. But now we need to turn them into power. And that's where the Fire comes in. The fire to keep moving forward against the odds. To hang on and endure beyond the norm.

But how do we access that fire?

First, I guess, we understand that nobody is very brilliant or brave all the time. That wouldn't be real. It's about delivering that little bit extra when it really matters. In the big moments.

It's always been the unwritten motto of the British Special Forces: Always a little further.

These were some of the first words ever spoken to me when we finally passed that Selection course. I'll never forget being stood in an underground bunker on the Hereford base, trousers ripped, blood and mud smeared on our faces at the end of eleven long months, but our hearts were proud. And the Commanding Officer stood in front of us and said this:

'The difference between ordinary and extra-ordinary is often just that little word "extra". And what I am going to ask you to do now is not going to be ordinary, it's going to be extra-ordinary. But what's going to make the difference is, when not one ounce of you thinks you can, you're the ones who will turn around and give me that little bit extra. That's what makes our work here extra-ordinary. Always a little further.'

Words that as a young soldier meant the world to me. To know that whenever it is grim, and people are throwing in the towel, you make that a trigger to give more, when most people give up.

But sometimes you've got to dig deep to access that fire. But if you dig, you'll find it. The fire is always there. Sometimes in life it gets a little hidden, sometimes it's just an ember. But it's always there, and the fire can change everything.

When I was on Everest, the mountain claimed four climbers' lives. Two died of the cold and two fell. It's what high mountains do: they kick the hell out of you. They ask that question: What are you made of when it hurts?

And I'll never forget the final hours on Everest. We were at 28,000ft, and we'd been climbing now for over fifty-five days. It's minus 40°C, and the dark ice face of the final 1,000ft of the peak looms above. And I'm terrified.

The next twenty-four hours is going to change my life. I know that. But after so long at high altitude, I'm weak – really weak now – and mentally exhausted.

I'm in chest-deep powder snow in a steep couloir and it's never-ending. I keep sliding back with every step and I'm no longer sure I can do this. I really am not.

And this voice inside keeps telling me this:

'You don't belong here.'

'Give up.'

And I'm slowly bowing to the voice.

But remember: sometimes an ember is all it takes.

So I keep going . . . Just Never Give Up. NGU.

This fire's been my most valuable weapon. Not talent, not skills, not knowledge, but resilience.

If all you take from this talk is this: know the power of unrelenting, unwavering resilience.

Because it's the storms of life that make us strong, and as with all dark nights, sometimes we just have to hang on in there. Doggedly. The dawn will always come. The light will always win. And the fire, in nature, is all powerful.

65

FAITH MATTERS

So, we have the failures, the fears, the fire. And finally there is Faith. Faith in yourself. Faith in others. Faith in the universal force of goodness.

And at 28,500ft on Everest, still 600ft below the summit, for the first time since that rehabilitation hospital I begin to sense this other voice inside me. A better voice, willing me on.

'I'm with you . . . lean on me. Keep going, forward, falling. Get back up, the summit will come.'

We all face our 'Everests'. Whether it's an actual mountain, an injury or an illness, or simply holding down a job and raising a family as best we can. Because Everest really is just a state of mind.

I finally reached the summit of my Everest.

And it was extra-ordinary.

You see the curvature of the Earth at the edges and you're aware you're truly somewhere special. The wind finally dropped and I collapsed to my knees. On that mountain, I experienced the voice of doubt being silenced, drowned out by something better, something that's hard to describe . . .

But what I do know is that I've leant on my Christian faith in the grimmest of moments. This universal presence, like a secret strength and backbone. And I still draw on it every day.

My father was alive when I finally reached the summit of Everest, but he died soon after. But not before I'd got to say to him, the man who had taught me to climb as a young, shy boy, that he'd climbed that mountain with me, every faltering step of the way, together.

One of the only things I took from the summit was a scoop of snow in a little bottle. And my father used to take a sip of that water every now and again. Like our magic potion. For strength.

But here's the thing. It was never about the water.

It was about understanding that achieving dreams will mean embracing Failure, and that Fear will be a constant journeyman. It's about understanding that sometimes you've got to dig deeper than you'll ever imagine, but we're given a Fire with which to fight, and a Faith with which to endure . . . Just Never Give Up.

That's really what my late father, that wonderful man, taught me.

So I guess what I'm trying to say through all this is simple. First, that I'm no hero. Truly. I struggle often and I'm no stranger to crippling doubt and fear. But I know the weapons that have served me best. And they come from within. Not without.

Those four Fs: Failure . . . Fear . . . Fire . . . and Faith. They mean everything.

66

TRUE WEALTH

But there's one final element to this talk, because reaching our summits should never be our whole story. This final element is about true wealth: it's about being grateful and about being kind.

I've climbed mountains with men who have lost their legs in war and yet they never stopped smiling. Look at all we have. Gratitude. Because all of us stand on the shoulders of some giants.

I do. My late father, those who stood beside me in the military, on Everest, on many expeditions since. The camera teams on our TV shows – the unsung heroes – who work so hard against crazy odds.

It's about seeking humility. Knowing our place in the universe. Gratitude for the good stuff.

I can count a solid twenty-one times I should've died during the many early episodes of our TV shows. Not a number I'm proud of, but in the early days we didn't know any better.

Bitten by snakes, falling down crevasses, pinned in rapids, avalanches, rockfalls . . . that time in the Costa Rican jungle I thought it was a good idea to use a vine to rappel down a 100ft waterfall. That vine wasn't as strong as I'd anticipated!

It all taught me the simple lesson: Don't be an idiot, and always be grateful for life.

Then, secondly, be kind. It's probably the most powerful thing I get to see in my role as Chief Scout. The way kindness changes people.

We now have fifty million young Scouts. A worldwide force for good. All having taken a Promise: to be kind and helpful. And when I see young Scouts holding the hands of dementia sufferers in senior citizen homes, it's a beautiful reminder of how kindness can better our world.

So treasure those around you: your family, your friends. Don't take life – or them – for granted. I do, all too often. But our real wealth is always going to be found in our relationships. Truth.

I hope that somewhere in all this are some little nuggets for you. It's about the simple things that keep us moving through the dark nights that we all face from time to time, and that keep us smiling in the brighter days.

So, remember: You're made amazing . . . Stand tall . . . and NGU – Never Give Up.

BG EXITS STAGE LEFT

67

THE CONSTANT BATTLEGROUND

FOR SOME PEOPLE, the adrenalin of performing on stage is a buzz that fuels them. I've never been like that. Instead, I am always just relieved to be done. And normally pretty drained by the time I get off stage.

The travel time often settles me, and gives me a chance to reflect about stuff. How strange life can be at times, but it is also a battleground of sorts. For everyone.

I have learnt that few successes ever come easy. Success is the product of struggle. The fruit for those that endure through the failures.

In all our lives, there will be some beautiful moments, and maybe some incredible summits along the way, but generally life is a struggle. However much fancy stuff and cars and money you might have, the truth is, we're all just trying to keep this thing called life on the rails. As best we can.

When I see outwardly perfect and 'successful' people, and it all looks effortless and easy, I no longer believe it. And I don't really like it. I know it's not real. I, for one, am just hanging on in there. Doing my best. Day to day. And often it feels like the wheels are going to come off at any moment. That's just life. I understand that now.

My point is, please don't think I have made it. I haven't. And that talk you just read should be nothing more than a reminder of this – and of the weapons that serve us best as we pursue our dreams and endure the battles along the way.

I have always known that true success can never be measured by the size of an audience or bank balance. To measure success like that is a recipe for unhappiness. The number is never enough. It's a hole we cannot fill or satisfy.

True success, to me, can only be measured by the quality of our relationships.

Screw those up, and whatever they say, you're poor.

68

WARRIORS BESIDE ME

SOMEBODY ONCE SAID: 'Show me your friends and I'll tell you your character.' I've always felt that to be true. In fact, I would argue that one's closest, trusted friends are the best measure of us all.

When my father died, I felt lost in so many ways. But above all, I missed a mentor and friend to talk through stuff with. You know the sort of person I mean.

Nicky and Pips Gumbel are among the kindest friends I have. They run an incredible, warm, open, hospitable church in central London. I originally met Nicky at our local gym, so to me he was always a work-out and squash buddy rather than a vicar. Above all, we laugh a lot, and I always love to tell him irreverent jokes, with the suggestion he uses them in his next sermon. Thank God he never does.

There is such power in trusted, wise words from close friends about which direction to take at key junctions in life. And so often, our lives become a product of those decisions. Both Nicky and Pips have been that guide to me, many times. In the highs and through the lows. Always constant, always kind, always forgiving, and invariably right. I value my friendship with both of them more and more as the years go by.

Then there's Jim Hawkins, who I have known since I was sixteen. Many years ago we made a commitment to email each other every single day for the rest of our lives. I had half suggested it as a joke. He took it literally. We haven't yet failed to complete a day without sending each other a note of encouragement for the day ahead. I'm pretty proud of that.

We use a simple life app that gives us a daily reading and we pick a bit out that feels relevant to what's going on. This routine has become such a positive thing in my life. It's a safe place to prepare for the day ahead, to share any struggles and fears, as well as a platform for giving each other honest, funny, vulnerable encouragement through the storms. In truth, it is my church.

Doing this every single day isn't always easy. In fact, often it is really logistically difficult, especially when out of comms or filming from 4 a.m. until late into the night in some jungle. And sometimes it is just difficult to find the motivation to carve out those ten minutes when you are tired and maybe not in the mood. But we do it. And it has become one of the pillars of my day.

The other key influencer in my life, and a constant friend through the years, has been my best man, Charlie Mackesy. It's fun for me to write that name nowadays as for twenty-five years he has been a best friend who was known to many of my other friends and family as the dirty, scruffy artist who stays with Bear and Shara all the time with his dogs.

That changed last year when he wrote an amazing book that catapulted him to the top of bestseller lists around the world, establishing him firmly as one of the most successful authors on the planet. Not bad for a man who last bathed in the early 1990s.

Charlie has been a friend to Shara and me since before we were married. He has been a rock in our lives, endlessly kind, fun and loyal. We continue to this day to swim naked in muddy rivers in winter, and to drink tea late into the night in front of open fires. Those sorts of friendships become ever more beautiful.

I include in this list the likes of Mick Crosthwaite (from Everest, school and the SAS), Trucker Goodwin-Hudson (with whom I went through Selection for 21 SAS), Hugo Mackenzie-Smith (school friend and buddy through many misadventures), Gilo Cardozo (from our paramotor Mission Everest and so many other trips), Nige Thompson (from our Arctic RIB expedition) and Al Vere Nicoll (another school friend). Best friends, and old friends. From good times, and some difficult times too. The best of people to share the adventure of life beside.

And as our respective families have grown, I love how those adventures adapt and continue. I love you guys very much.

Then there is Verity Allinson, who has been a constant source of support, friendship and kindness in Shara's and my life, especially with our boys, growing up. From day one, when she arrived as a helper to Shara – and left her scarf and handbag behind in a fluster – to having become one of the best, most diligent, hard-working people in our lives. On so many levels, Verity is an inspiration.

My editor suggested that this chapter should really be in the acknowledgements section, but these friendships are such a part of my life that they can't be sidelined. I want them sealed in here, safe, embedded at the core of my story. It is where they belong.

I recently filmed with Michael McIntyre as his 'celebrity guest' on his Saturday night BBC chat show. The 'Send to All' segment involved him texting all our contacts with a spoof message and then reading out some of their replies. It was all pretty funny. But after we had wrapped and were saying goodbye backstage, he took me aside and said this: 'I've done this Send to All routine on so many people over the years, but reading all those replies from your friends in particular has been amazing. And what is so lovely, and so apparent in your and Shara's life, is that you predominantly have friends that have been beside you from the beginning. In this business, that says a lot. But what was so clear, above everything, is that you have so many people who so clearly have your backs.'

It was the greatest compliment he could pay us.

69

FRIENDS ALL OVER THE WORLD

IF YOU HAD to ask me what fame does, I would say that it tends to amplify whatever already exists in your life. The good, the bad. The ability to make things happen. The doubts. And those fears.

It all simply gets bigger.

As a family, for example, it has meant that when we are out in public, I sometimes get shared. That's all. It's not a big deal. Our boys get it. They will always step aside (despite people's best intentions to want to include them as well), and just let the person take the selfie. For me, it's simple: I smile, squeeze their shoulder, and move on.

Moving on. It's a family rule: always be warm, be nice, hear the person's story (and there almost always is one, and those stories can be amazing sometimes, by the way) but then move on. Always forward. Like a firefight. (But for the record, I truly have some of the friendliest, most inspiring fans anyone could ever hope for. And I never take that for granted. The connection is a privilege.)

Then there's the whole thing of children in the media or spotlight.

For us as a family, we have just always had a rule to try to keep them out of the public eye as much as possible. Once they are sixteen it's up to them, but while they are young, we want to keep them away from any spotlight as much as we can. Life has enough pressures for kids growing up as it is without the whole recognition thing.

It's amazing how having the family close around me, or holding hands with Shara or the boys, often means fewer people will come up for a selfie. I don't know why it's like that but as soon as Shara goes into a shop or I am suddenly left solo, then that's when life gets busier. It's a family joke now: 'Don't leave Papa.'

I like that sentiment.

Final words on celebrity, before we move on to way more interesting stuff, is that I always try to see recognition as a good thing, a blessing. I often say to our boys that it means we simply have friends all over the world. And that is an amazing truth.

People want to shake your hand and tell you their story, whether they have been camping with their daughter or climbing a hill with their college buddies – whatever it is. I never mind that. It's a good thing, and I love that sense of connection and community that adventure and a TV set can bring.

After all, friendships make the world go round.

It has taken me a while to get used to any level of recognition, and to see it for what it is. When it first happened, I struggled with it. I am not a natural extrovert. From the beginning, I have struggled with the spotlight and had never aimed to be on TV. Those things simply came as a by-product of seeking out great adventures.

It is worth noting that in my experience, 99 per cent of people are respectful, discreet and encouraging. But I always saw recognition as the downside, the price of the fun. I considered it something to dodge. But with time, I have learnt to see a better way – that, in truth, it isn't about me. It's about sharing a moment of warmth, and trying to leave that person with a sense of fun, light and positivity. And when I figured that one out, then it all became much less hectic, scary and unwelcome.

Finally, and most importantly, is realizing that this whole fame stuff isn't real life. It's transient and mainly unjustified. And there will come a time when no one will care. That is why in life you have to invest time and love and energy in the relationships closest to you. So that when the fluff is gone (your title, your work and any sort of identity others might give you), you make sure you are left with something real – something that you have invested in and protected.

That's why *family first* is such a key principle to me.

The lesson is simple: smile, try and spread good energy . . . and always keep moving.

70

FAITH TO MOVE MOUNTAINS

NO DOUBT, FAITH is really hard to articulate. It is intimate and it is fragile. And it is even harder to share openly. But if a book like this is to be honest, then it should have all those elements in it. The fragile and the intimate.

To me, faith and religion are poles apart.

In fact, it always seems that the great men and women of faith, throughout history, have been the ones who have rallied against religion, yet promoted love, courage and kindness.

Look at John the Baptist, one of my great heroes. He lived outside, in the desert and in the mountains, was the best friend of Jesus, ate honey and locusts, and had no care for his own glory. Totally free, and wild at heart. He rarely ever went to church, and got mad with the hypocrites and religious folk. All he wanted to do was point people to the 'light of the world', then go bathe in the river. Legend.

I was always intrigued by what Jesus made of 'religion'. Sure enough, He was pretty clear.

You brood of vipers. You blind people to the truth.

So what is this truth?

That's some question.

And I am far too flawed and lacking in understanding to know the full answer. But I love the words that Christ spoke to the broken and the outcasts.

I have come to bring life.

Words to live by.

If I had to sum up my journey of faith, and what it means to me, and what it does, I would say it like this. I wrote these words in response to a really sick teenager in hospital, who asked me what my faith is really about. He was facing a frightening and uncertain future. I wanted to answer him properly. It took me a while to articulate. This is what I wrote:

Hey buddy – here is my best shot to share my faith.

Take whatever part you like, or leave it all. But I wanted to try and answer your question as best I can. And as honestly as I can.

You're going to be OK. And I am here for you.

So, here we go:

Some people call faith a crutch. But what does a crutch do?

It helps us stand and it gives us a weapon to fight with. And as time goes on, there's no doubt that I need that strength inside me more and more, every day.

When it comes to quietly bowing the knee and asking for God's presence to bring me peace, to strengthen my spirit and to lead me into light – well, I have nothing to lose, and everything to gain.

Faith, to me, has been the great empowerment in my life. Quiet, strong, personal . . . and it has never let me down.

There have been times of much doubt along the way – for sure – but faith and doubt are simply two sides of the same coin. Doubts are just part of life. I just try to keep focused on the good stuff.

A few times I have walked away from faith completely. And I've survived, for a while – but, alone, I somehow never feel fully alive. Time, and many adventures, have taught me that I need the life-giving presence that Christ provides.

Faith is a point of awareness. And I believe it to be the starting place of all true adventure.

So often, faith in a higher power or God, or whatever we call it, has been strength to a failing body and light to a dark path.

That presence is much more than a crutch. It's a backbone.

As I once saw written on an old wooden cross in a small mountain chapel: 'Christ beside me. Christ within me. Christ to shield me. Christ to win me.'

Sometimes, I hold on to just that thought alone, and it helps me through the day.

71

YOU VS. WILD

I SOMETIMES FEEL that the bigger the project, the bigger the opposition that stands in its way. And often this opposition comes in strange forms.

Sometimes it is from my wider family being a bit tricky, or maybe it is from some rogue curveball that gets thrown our way, an untrue or unkind series of press articles or a difficult situation with one of our wider team. But it is there. Still the same.

I have seen this pattern in my life often. Battle and blessing. I now take the battles as a reminder that we are doing something right. Something big. Blessing is coming. But first we have to deal with the storm.

The day we were due to start one of the most groundbreaking new TV series we had ever taken on was a big moment for us. Netflix had approached us to do a large-scale adventure series, not dissimilar to *Man vs. Wild*, but with a big ask, and a big question attached. Did we reckon we could make the show interactive for viewers? Could we do the journeys in a way that Netflix viewers could make their own decisions about which way I went – what I did at every turn in the jungle, across a desert or on a mountain? Choose your own adventure . . . on steroids.

We would call it *You vs. Wild*. And it would be the first ever interactive adventure show in history.

No pressure.

The planning was at best intense and at times chaotic. The mapping of all the choices, and the repercussions of endless different decisions, meant that the episode documents started to resemble a snake's wedding of options and directions.

Luckily, our production team of Rob Buchta and Ben Simms – both veterans of *Running Wild* with me – dug in, and kept unravelling the puzzle. It took some serious brain teasing and conundrum solving just to get us to 'day one' filming.

Soon, the time came. I was happy because the show was kicking off near home, in the Snowdonia mountains of North Wales. It somehow felt fitting.

The day before, though, I had started to come down with a really bad fever. I was struggling to stand up without feeling faint. By the time the chopper landed to fly me to production, I was really doubting whether I could do this. We had so much on the line here, and so much work had already gone in to get us to this starting point. Yet I felt horrendous.

I took a deep breath and climbed aboard the helicopter, a little voice inside telling me all would be OK.

Par for the course. Battle and blessing. Hold tight and muscle through. This show is meant to happen.

Del was a trooper and helped me load up with meds, and at dawn the next day I dragged myself from my bed, weak and sweating, and went up the mountain to work.

It was a testament to our crew that we pulled off that first shoot. They carried me. Totally. That's what great teams do: they take up the slack and help you stand when you are struggling. We even got one of our team to double for me for second takes of shots, and in between filming I would collapse in a heap and rest.

It really was a case of stumbling through that first shoot, and it was the sort of action that needed energy and commitment. Backflips from helicopters into freezing lakes, rock climbs, crawling through dank tunnels and caves, and chest-deep river crossings. It rained incessantly, was sub-zero, and the crew were having to shoot in a whole different style, while also keeping track of where the hell we were in the story.

The crew saved me, saved the day and saved the show. I really felt that if we hadn't done that first week and started when we did, Netflix might have come to their senses and realized how overly ambitious we were being. It would have been very easy to delay the project a year – and at that point who knows what might have happened?

You vs. Wild was the closest any of us had come to making a complex movie, with pre-planned action, stunts and a clear story narrative. By the time Netflix started to see rushes and the inner workings of what we were shooting, combined with an appreciation of the technology needed from the Netflix platform, it became apparent we were all winging it. Throw in multiple viewing devices and languages and Netflix territories, and the scale of the project was becoming clear to all of us.

This was why interactivity in media had hardly ever been attempted. It's one thing in a studio, where everything is sanitized and calm, but trying to do interactive adventure and technology in a jungle, during a torrential downpour, was hard. In the words of one Netflix exec: 'It feels like we are building this aeroplane while we are airborne.'

I stumbled back after another day in the mountains and collapsed in a heap. I was beat. But based on the battle and blessing theory, this series was destined to make a splash.

I set my alarm for 5.30 a.m. and prayed I would feel a little stronger the next day.

72

KEEP INNOVATING

BY THE TIME we wrapped up *You vs. Wild* for Netflix we had travelled all over the world and had some epic adventures and unforgettable moments – but it was the edit that was the real monster for our team. Somehow, they pulled off the impossible, and by the time of the launch we were all fired up to see how people would react to the first ever interactive adventure show.

The response to that show was amazing to see. It received over a hundred million downloads and was one of the most successful non-scripted shows of the year across all of Netflix. It got Emmy nominated and went on to win the award for the best multi-platform TV format.

As Scott, one of our core safety team, observed recently: 'When I used to go and speak at schools about the BG Survival Academies, I would always open up with: "Who here loves *Man vs. Wild*?" And there would always be a roar of excitement. Nowadays, when I say *Man vs. Wild* to young kids, they look blank. Then I say: "Who here loves *You vs. Wild*?" and then they go really crazy.'

And that's life. We turn up, do our best, and if we're lucky, we might make something that touches people and inspires them – but everything has its time and place, and nothing stands still. The successes or failures come and go. Nothing lasts for ever. That's why we have to keep innovating if we are to keep inspiring. Always try something new. Take risks. Fail. Fall down. Go again.

This is the beating heart of the message of all our shows, and none more so than *You vs. Wild*. For me, the great privilege of making the first ever interactive adventure series was to be able to introduce a whole new generation to the thrill and excitement of the great outdoors, all the while helping kids to be able to assess and make smart choices under pressure.

Netflix have proved to be great partners for Del and me at The Natural Studios, and pretty soon we were into pre-production for the first of two movie versions of *You vs. Wild*. These will be our first ever feature films and an exciting next step for sure.

It was June, and school holidays were looming. Time to get away to our island in Wales. But just before we were due to go, I was asked by the CEO of Netflix, Ted Sarandos, if I would help him at the company's annual convention in Iceland by being interviewed by him on stage.

It was an honour to be asked, and great recognition of the power of *You vs. Wild* among the Netflix family. One final engagement before the summer. I said I'd do it, and the plan was to bring Jesse along for the ride.

Ted sent his personal jet to collect us, and Jesse and I headed north to the land of fire and ice. Just being on Ted's jet with Jesse was such a reminder of the journey we had been on as a family. From simple beginnings, through so many failures, stumbles and mistakes, we had persevered and kept our focus. We had built a great team and stayed true to the vision. I wanted Jesse to see how much those things matter in life, and for us to be able to share the moment. I wanted to say to my family, look at how far we've come and what we have built, together – with much good fortune along the way. Something truly special.

Before I knew it, I was backstage at the convention centre with Jesse and Ted Sarandos, ready to speak to thousands of his key staff from around the world. I was nervous. I always am. But how much braver I felt with Jesse beside me. All my family have that effect on me. With them beside me, I know I can do it.

For one of the most powerful men in media, Ted Sarandos is such a humble, family-centred guy, and on stage he was so kind and encouraging to me. I felt very humbled even to be there. We laughed and had fun, and I told some of the stories behind *Man vs. Wild*, *Running Wild* and, of course, *You vs. Wild*. It felt good to be a part of the Netflix family. Keep with the movers and shakers. Keep innovating. Always be grateful.

That evening, I walked the streets of Reykjavik with Jesse, just us two. Netflix had kindly booked us into the best restaurant in the city, but I had politely turned it down. I knew what I wanted to do instead.

We walked down to the harbour and found a fish and chips van, near to where the fishing boats came in to dock. It was a quiet evening at the dockside but the smell of fresh fish and crab was rich in the air. Jesse and I sat there, fish and chips wrapped in paper on our laps, looking out across the water.

I pointed out to Jesse the exact spot where our little RIB had pulled into this port some fifteen years earlier, after our crossing of the North Atlantic Ocean. We were five pretty broken men that night, having barely endured that monster storm some 500 miles offshore, surrounded by ice and wild seas that had constantly threatened to overturn our small boat.

All our communications and electronics had gone down, but our single engine had kept going, despite the pounding we'd been getting. By the time we limped into port we were like zombies, having not slept for many days. We were all close to hypothermia, we had frostnip in our feet and fingers, and we knew we were very lucky souls to be alive. I would never forget the smell of the docks on arrival. The stench of fish was the sweet smell of survival.

To sit there and look across the same harbour, with the same sights and smells, but this time with Jesse beside me, was special. He had been only a matter of weeks old when I had set out on that Arctic expedition – and it had almost killed us. What a fool I had been even to risk that. But how much it meant to me now, to be there with him, aged sixteen, sat eating fish and chips together. A beautiful moment.

Life can be so circular sometimes, and sharing those 'come around' moments with those we love is always good. Plus, the fact that we also had the Netflix jet on standby to fly us home the next morning was also nice.

When we arrived in Iceland all those years ago, we had been helped by a local Icelandic mechanic who had been following our journey online. Bogi had done so much for us at that time, finding us beds to sleep in, helping us with repairs, food and fuel, and driving us all over the place as we prepared for our final leg south, to the Faroe Islands and then Scotland. Bogi had asked for nothing in return.

I so wanted to find Bogi again, to thank him and give him a hug, and also to introduce him to Jesse. I'd tried a whole bunch of ways to locate him, through social media and beyond, but there are many Bogis in Iceland and without his surname I had failed to find him.

I doubt Bogi would have wanted any fuss or attention anyway. But if you're reading this, Bogi, thank you. It was all worth surviving for.

73

AMAZON: WORLD'S TOUGHEST RACE

I'VE ALWAYS LIKED the notion of constant and never-ending improvement, and trying to stay one step ahead of the competition – *always that little further*.

It is why we are constantly developing new shows, new formats. Some work, some don't – but no one can accuse us of not aiming high.

Last year, we partnered to make a really special series called *Hostile Planet*, for National Geographic worldwide. It took over three hundred days of wildlife filming for the crew to put the series together, and the result was immense, showing wildlife at both its most vulnerable but also its most resilient.

From Himalayan snow leopards hunting, and falling and surviving 200ft drops off cliffs, to Arctic polar bears, in desperation, resorting to hunting whales for their survival, *Hostile Planet* showed how climate change is affecting the lifestyles and behaviour patterns of animals all over the world. As the planet gets hotter, and the weather less predictable, as storms become more ferocious, with droughts lasting longer, and flooding and forest fires more frequent, the life of wild animals, already on the edge, gets even more precarious.

I was so proud to host this series, and to follow the likes of the BBC's *Blue Planet* and other landmark natural history shows that historically have been fronted by the wildlife legend Sir David Attenborough.

David has without doubt inspired a generation of young people to care about the future of the planet, and its wildlife, in a way that is desperately needed. All of our team saw first hand the devastating effects these extremes of weather are now pressing upon the wildlife of our world. It is going to be down to you and me, to all of us, to keep fighting this fight, to insist that governments take bold action to curtail plastic use, fossil fuel dependence, and to protect our wild places and the creatures that inhabit them.

Hostile Planet championed that message in the best way possible: by showing what happens to animals when we don't. The show was shocking at times, but also so uplifting, and it got four Emmy nominations for the effort, which was due recognition for all that the team went through to put the show together. But the series didn't return, sadly. It had proved simply too expensive and too time-consuming to shoot to make it sustainable for a commercial channel like Nat Geo. But to have dipped my toe into the natural history waters was really important to me, and who knows where that one will lead?

One show, though, that I knew would be big was bringing back the original *Eco Challenge* adventure racing format that had last aired on TV almost twenty years previously.

Mark Burnett is the dynamo British producer behind some of the biggest hits in TV history, such as *The Apprentice*, *Survivor*, *The Voice* and *Shark Tank*. He's a former British Paratrooper, and we have known each other for many years, initially as Brits working in America. He had since sold his media business and was heading up the legendary MGM Studios. His heart, though, had always been with one original show that he created, the one that launched his titanic career: *Eco Challenge*, the toughest adventure race on the planet.

When Mark reached out to me to partner on bringing the show back to TV, it was an honour. And the timing for both of us was perfect. Mark is also a man of faith, and whenever we had seen each other over the years he had always let me know he had my back if ever it was needed, and that he was praying for me to succeed in all I did. True support and kindness like this are all too rare in the glitz and glamour of Hollywood, and I always remembered it.

True to his word, Mark reached out, and MGM and Del got to work together to find the perfect network partner for us to sell the show to. We would rename it simply *World's Toughest Race*.

Together, Mark, the original *Eco Challenge* team and I set out four clear conditions. First, we had to stay true to the original series: *Eco Challenge* must always remain an expedition with a stopwatch. Second, the environmental factor had to stay front and centre. This included every competitor doing all they could to protect and preserve the wildernesses they were racing through, even to the extent of carrying out their own faeces with them in their backpacks. *Leave only footprints, take only memories.* (And SD cards.)

Third, each team must stay as a complete team, or they're out. You finish together, or not at all. And finally, we must remain the world's toughest race. Always. And in this regard we set out to deliver.

The actual adventure course that our race organizers, Kevin Hodder and Scott Flavelle, along with Lisa Hennessy and the team, put together was immense: 672km of wild mountains, jungle, swamps, canyons, rivers and oceans. Using multiple adventure disciplines to navigate the huge distances, the sixty-six international teams of four racers each would be sleeping barely an hour or so a night, over eleven days, in an attempt to complete the course.

No one had ever staged a race of this complexity or ambition. Not even close. No pressure, no diamonds.

To run any adventure race is a big and expensive business. It is why most adventure race companies fail: you simply can't sell enough tickets to cover the costs of running a truly groundbreaking race, particularly a global one. With the right TV partner behind us, though, we knew we could do what no one else had managed. But it would be expensive.

Not only was Mark great friends with Jeff Bezos, the founder and CEO of Amazon, and the world's richest man, but our team had also taken Jeff and his board away on an adventure some years earlier.

Amazon had reached out to our Bear Grylls Survival Academy and asked us to put a survival scenario together for Jeff and his senior team as part of their corporate away break. I like to think we delivered. We had them climbing cliffs, dodging rattlesnakes and even eating maggots – which Jeff was determined to lead the way on. And good for him. As the world's wealthiest man it must be tempting to stop doing the uncomfortable, and change your lifestyle to all caviar and champagne.

In fact, conversely, he had written to me afterwards, saying:

> Thank you, Bear – we had an absolutely great time with your crew. They had us laughing out loud continuously. In addition to all the laughter, I also learned that grubs are delicious and earthworms less so. Jeff.

The friendship between Mark and Jeff, and the trust that Amazon Prime had in us to deliver, was a great combination. We were given one of the biggest TV budgets in Amazon Prime history to go and create something unique and inspiring. To do that meant bringing in the best. It's a principle both Mark and I have always lived by. Don't skimp on hiring great people, especially in terms of safety.

With some of the world's best mountain guides and pilots – not to mention a total crew of racers, support teams, paramedics, camera crew and volunteers of just under a thousand people – this show was on a scale unlike anything we had done before.

Mark wrote me a short text to say simply: 'I know our partnership is meant to be. To bring positivity and that adventure spirit to millions. I am so proud I have handed over the reins to you to produce and host this. This show is going to inspire the world! MB.'

That's friendship, and that's leadership. Empower others. The next generation. Then give them the tools to deliver.

On *World's Toughest Race*, we witnessed teams from every corner of the world battling and giving their all for their dreams. Teams of military veterans with missing limbs, teams of cancer survivors, young Indian teams, the first African American adventurer racing teams. You name it. Stories of tenacity, courage and friendship that left me at times in tears.

When the show aired it hit a chord with so many people. It was the story of unsung heroes battling adversity and winning through. It was never about the medals but about the hearts of those who gave so much and helped each other across that distant finish line. By the end, the series was Amazon Prime's most viewed show of the summer. Across movies and all of TV.

The message of the show was simple: together we can do so much. It was a message of unity, hope and inspiration. Strong values, which resonated with viewers as well as the team at Amazon, who had always believed in us.

Whether or not we go on to do more seasons of *World's Toughest Race*, only time will tell (although there is no doubt it was an expensive series, even for the mighty Amazon), but either way, I feel certain that we're going to do much more with the company in the years ahead.

In fact, only today I got an email to say that their corporate team want to partner with us on our global education initiative, BecomingX. They want to help sponsor thousands of underprivileged kids and schools to give them access to our BecomingX resources, which will help those youngsters get a leg-up in life, in terms of skills, knowledge, attitude and relationships.

I love that. A company putting their money where their mouth is. On one hand they enable us to make inspiring and motivating content for their media channels, then they support us to give tangible, educational deliverables to kids worldwide. That's how we make the world a little better. Partner with the best, look beyond ourselves, and know that, as with everything, what we sow is what we reap.

74

KEEPING YOUNG

IN BETWEEN OUR main shows, we had also shot a series for ITV and their kids channel, CITV, based upon the success of our Survival Academy.

The show was called *Bear Grylls Survival School*, and we got to take some young teenagers from all walks of life to the mountains of Wales, and put them through an incredibly demanding two-week survival adventure bootcamp. It was about getting them off their tech and learning to depend on each other in some big moments. Dealing with the worst that Mother Nature could throw at them and working together to overcome some huge challenges that we threw at them.

People thought us crazy – the kids would never survive with no phones, let alone having to endure daily PT sessions, swimming in cold rivers, eating cold rations, and having wet gear and limited sleep. They would surely crack. But they didn't.

I remember saying to them on day one that we were going to treat them like adults – so don't let me down. Again, they didn't. Their journey was so inspiring. It was amazing to watch them grow personally through the hardships, and to see how they learnt to respect each other through the adversity.

By the end, I had tears in my eyes as I stood in the rain, at the foot of this bleak Welsh mountain, flanked by our small team of former

Royal Marines instructors. There and then I awarded them their black BG karabiners as symbols of friendship, risk and strength. They had done us all so proud.

The TV show flew in the ITV ratings, more than any of us had ever hoped for, and then the series aired on Netflix as well. I am so proud of making TV that empowers even the youngest of demographics, and rebukes those who think teenagers are a lost cause.

What this taught me is that we had a huge opportunity to inspire a fresh generation of young people to 'go for it' in their lives and to embrace the power of adventure. But maybe there was a smarter way we could do this.

One factor that was becoming ever clearer is that, however good we try and make our TV shows, they will always be limited in two ways: the number of episodes we can produce and the ever-increasing odds in terms of risk and danger. However safe we try to be, I will get older, and the odds of injury will increase – not to mention the odds of some other global disaster, outside of our control, that could potentially limit how much we could film.

In fact, there are a thousand things that could shut production down, and that makes our future inherently vulnerable. That is why I have always known that a really smart part of our BGV world to develop and build for the future would be a 'young Bear Grylls' animation. Whether for TV or film.

The smart part is that we could introduce everything that people loved in the original *Man vs. Wild* show to a whole new generation of young people without travelling anywhere or risking life and limb every day. After all, animation characters don't get injured and can do hundreds of episodes a year.

Step one was finding the right team. It took a few attempts, like all good things, but finally we found the right partners in Platinum Films, based at Pinewood Studios – home to the legendary 007 sound stage and so much more.

Nigel Stone, their CEO, had an unshakeable belief from the beginning in the huge potential of a 'Bear Grylls Young Adventurer' animation as a vehicle for inspiring kids with great values in different countries and languages worldwide. Some things like kindness, bravery, loyalty and resilience are universally needed and admired, but rarely shown in a fun, adventurous, aspirational way.

We knew we could build hundreds of positive adventures around a young me and a group of buddies getting into all sorts of adventures and scrapes, protecting the environment, dodging school bullies and leaping out of airships over jungles.

We currently have the first of three animation movies in production with Bron Studios. The first drawings are in, the voiceovers are being laid, the music score is being written – it's going to be fun.

Maybe they will fail. Maybe the films will be huge. That's the adventure of life. But nothing ventured, nothing gained.

75

BE PREPARED

I'LL NEVER FORGET the day when the UK Scout Association asked whether I would consider interviewing to become their youngest ever Chief Scout. I look back on that moment as one of the greatest privileges in my life.

Founded in 1908 in the UK, Scouting has grown into a worldwide organization some fifty million strong. A symbol of peace and unity, it is now the greatest youth movement on the planet. Rooted in friendships, family, respect and courage, Scouting symbolizes so much of what I believe in.

It was an honour even to be considered as their Chief. Especially as I was such an unassuming, nervous Scout as a kid. In fact, I was probably the quietest, shyest kid you could ever imagine, and you certainly wouldn't have noticed me in the Scout hut where I first joined as a wide-eyed eight-year-old.

I hardly earned any badges as a kid in the Scouts, but I loved the spirit of it all. That sense of adventure and excitement. Putting on my uniform, being taught real skills for life, such as how to cook a sausage with just one match. Formative moments.

I will never forget being sat on the floor with that one match and one raw sausage and trying to figure out how the match was ever going to last long enough to cook the thing. Then when a more experienced Scout taught me that I had to use the one match to light a fire, which I could then cook the sausage over, it was a light-bulb moment. *Of course. How stupid could I be?* I told you I wasn't a very brilliant Scout.

But maybe that was the point. As they say in life, you don't want to peak too early. And certainly not at school or Scouts. Those places should be seen as our training grounds, where we try, and we fail, and we learn to get back up. Our young years should be where we build our resilience and thirst for life. Those years shouldn't be your summit moments. Those will come.

The part that made me most proud about being considered for Chief Scout was that the casting vote was from the young Scouts themselves, rather than the executive adult Scout committee – I loved that mentality. Empower the young people to make decisions that shape their future. After all, Scouting always has been, and always should be, driven by the vision and efforts of the young Scouts themselves.

But I wasn't the only person the committee and Scouts were considering as their new Chief. They interviewed many others. And then they interviewed me again. And again. In fact, I had never been interviewed so much for anything in my life. They kept telling me that it was the teenage Scouts who were fighting for me hardest. I felt that if they would fight for me, then I would fight for them in return – if I ever won the position. It made me ever more determined to get the job.

The day I was sworn in was a very humbling one. To follow in the footsteps of some iconic Chief Scouts of years gone by was definitely daunting, but I vowed to do my best, to do my duty to God and to the Queen, to help others and to keep the Scout law. I couldn't have been more proud.

Being Chief was initially a five-year tenure, but the Association then asked if I would extend my term. I have now been Chief Scout for almost twelve years. One of the longest ever, and definitely the youngest.

Since that first nervous day, standing on stage and promising that I would champion young Scouts across the world, it has been an incredible ride. I vowed to help young people from all walks of life, faith and gender, to enable them to experience some great adventures, and to help them build life skills and lasting friendships along the way.

Every day as Chief Scout is both humbling and inspiring. And it is all about the young people themselves. My heroes. I meet them all the time. Young Scouts who have maybe saved someone's life, or helped care for a terminally ill relative; or Scouts who have broken world records, helped vulnerable people in their community, volunteered in a global crisis; or Scout leaders who have quietly served their pack, for no reward except to be part of something special.

From the volunteers to the kids themselves, from the UK Scouting CEO Matt Hyde and the global secretary-general Ahmad Alhendawi to the youngest, most nervous new Cub making his promise and joining the Scouting family right now, as you read. Heroes all. And the power is that Scouting reaches the most vulnerable in our society and helps them find their way, and it inspires so many throughout their lives. It's about helping young people gain access to adventure, and to learn life skills that will help them far into their future. It's legacy stuff, and I love it.

Just today I was hearing news about the Scout volunteers in the Syrian refugee camps who are trusted ahead of many aid agencies. That necker, that scarf, stands for so much. It is a symbol of peace, unity and service that has immense power. And it is no surprise that so many world leaders and icons of sport, business and film, astronauts and Nobel Peace Prize winners, have at one time been Scouts. Scouting teaches young people about working together, about leadership, about our global responsibilities – and all at an age when we are so open to growth and learning.

Scouting changes lives for the better, it changes us all for the better – and I couldn't be more proud to be a small cog in this most incredible machine. Truly a worldwide force for good.

I have spent many days and weekends through the years travelling around the UK and meeting Scouts, as well as meeting so many international Scouts on the way in or out of filming, often in remote locations around the world. It reminds me that we are part of a global family. Bound together by positive values that endure.

So many times I have taken Shara and the boys with me on these visits, and we always come away at the end simply saying, 'Wow.'

I just have to look at our Queen's Scout day, which we run at Windsor Castle every year. It's the moment we award the Queen's Scouts their badges for having reached the highest rank within Scouts.

The Queen has generously allowed the Scouts to parade within the inner quadrangle at Windsor Castle for over fifty years, and there is a palpable sense of tradition and heart to that parade that is hard to describe.

Thousands of people, from every walk of life, all bound together by values of kindness and endeavour. Many of those young people have been on incredible personal journeys to get there, in every respect, and to see the tears and the look of pride on the faces of their families is so moving.

Over the years I have had the honour to guide many members of the royal family, including the Queen, the Duchess of Cambridge, Prince Charles and others, up and down the ranks of the Queen's Scout recipients, and almost invariably the royal host will comment to me about how the smiles and pride on the Scouts' faces are so powerful to see.

Experiences and friendships, hardships and adventures . . . they give us something that no amount of money or status ever can. Life sometimes does its best to put people down and button people up. Scouting does the opposite. That's the magic for me.

76

WORLD SCOUTING CHIEF

TEN YEARS AS the youngest ever Chief Scout was a record I never imagined holding – after all, we generally see ourselves as the least impressive manifestation of who we are, don't we? And in my mind I will always be that shy, nervous Scout aged eight rather than chief of anything.

It is a reminder to me that I truly owe a debt of thanks to those who have helped me and encouraged me, and often promoted me into positions that I would never have presumed for myself. That's the power of surrounding yourself with champions. And at Scouts we have many of those.

Collectively, as a team, what I do know is that we have made Scouting relevant again – in a way that is aspirational, and focused on friendships and adventure. In truth, it was always in there, but sometimes big organizations get bogged down in rules and regulations, and committees, and lose a little of the heart of what makes them great. In Scouting, that beating heart is adventure, learning cool skills and helping people.

When the World Organization of the Scout Movement (WOSM), which runs the global Scouting family across 169 countries, approached me to ask if I would take on the first ever appointment as Chief Ambassador to World Scouting, I was so honoured. I also knew it made sense to our wider mission, and I knew I could help them amplify the positive message of Scouting in countries where young people were crying out for pride, purpose and skills. Above all, young people want life skills that give them an edge in an ever more competitive world.

My first ever formal engagement as the Chief Ambassador was to speak at the United Nations in New York to many diplomats, UN folk and the global Scouting leadership team. I was pretty nervous, and it wasn't helped by the fact that I had flown in hot from a Central American jungle with a bout of diarrhoea after having eaten a particularly dodgy snake on a shoot.

I'd no sooner entered the UN building and got past security than the diarrhoea started to make itself known again. By the time I had muddled through a bunch of press interviews and was ready to take to the podium I was feeling pretty light-headed. The plus side of being in that sort of state is that you speak from the heart and you drop the airs and graces. I did that in spades that day.

I remember saying how we mustn't let Scouting, and its UN relationship, become too formal or too corporate.

'No, our role as Scouts – the greatest youth movement on Earth, a true force to be reckoned with, some fifty million strong – is to show the UN how we can make their global goals relevant and fun without being stuffy; to empower young people with a mission and pride; to lead the way in effecting change in governments, at national and international and local levels; changing the world for the better, inside out, to show how young people know best what we need to do in order to protect and unite our world and give our youth a future.'

I was really getting going. I saw our UK leadership team fist-pump at the back of the room. I took it a stage further.

'We need to be rule breakers and way more reckless . . .'

It was about now that I saw that same team start to grimace.

But I was too dehydrated and tired to worry. Although I do wince when I think back to how I told the seventy-two-year-old King of

Sweden that 'young people in his country love him for his playful spirit'. King Carl Gustaf, the UN's honoured guest and a founding board member of WOSM, looked at me a little surprised. I just hope he took it as a compliment.

I made it through the rest of the speech without causing too much collateral damage to any UN or royal figureheads, but my point was – and is – that we all have to change. Big organizations must always fight hard to stay nimble and relevant, and uncorporate. Or they die. And as Scouts, we must never lose our ability to rewrite old rules and conventions, to pioneer and take risks, to laugh and have fun, even when among global giants . . . to go always that little bit further.

The Scouting founder, Lord Baden-Powell, reportedly once said that if a hundred years from now Scouting was the same as it was when it started, then something would be very wrong. Yes, our values stay the same – classless, without ranks, and built on friendship, endeavour and service – but how we deliver that has to change.

We must continue to do more, in terms of evolving our practices, our symbols and uniforms, the ages, genders, promises and faiths that Scouting supports. Always adapting. Always listening. Always learning and growing.

In this sense, Baden-Powell got so much right, not to mention coming up with so many great quotes that I still use today. Such as his explanation of the Scout motto 'Be Prepared': 'It simply means we are always in a state of readiness, in mind and body, to do our duty.'

Or this one: 'Never say die 'til you're dead!'

I like that one best of all.

77

GLOBAL VILLAGES

OVER THE YEARS it has been amazing to attend so many Scout camps all over the world, and there is something so powerful about the collective force of goodness that emits from those gatherings. It's hard to describe.

I think it has something to do with volunteers helping, encouraging and giving time, care and energy to so many young people, creating an environment of selflessness that is amazing to be part of. Coupled with the raw enthusiasm of young people having fun, living their adventures and trying and learning so many new skills, all while being soaked through and caked in mud. It is no wonder you can hear the roar and laughter of Scout camps from miles out.

From the World Scout Jamboree in Sweden to the global gathering of fifty thousand Scouts in West Virginia, USA, I have seen first hand the power of Scouting to unite and inspire young people. But sometimes, for me, it is the small camps, where there are maybe only twenty kids, camping in some forest on the edge of a town, in a remote part of Australia or Belize or India, that remind me that what we have is so special.

There is a timelessness to those camps. Sometimes I have arrived and seen rows of old-style white teepee tents all lined up, with various flags fluttering in the wind beside them, fires burning and Scouts running around laughing, and that makes me realize that I could easily have stepped back in time to 1910.

Moments like these connect countries, they connect people, and they bring us all closer together through a common set of values that crosses borders and transcends language.

I hear so many stories from young Scouts, travelling the world, who have received such beautiful, kind hospitality from Scouts and their families in other countries. Young people and their families who have never met before but who, through Scout social media or old-fashioned letters, have offered to have Scouts to stay in their home. That's what happens when you have a common purpose, when you are part of a global family, and when you share great experiences.

These are all simply timeless truths, about how the great outdoors and the wild are universally loved and admired. How the natural world can both challenge and change us, how the night sky and the stars above can inspire us, how friendships, formed through adversity and through adventure, can last a lifetime. These are great things to know in life. And I never tire of them.

78

ONCE A MARINE, ALWAYS A MARINE

THE OTHER ROLE that I accepted during these years was to become Honorary Colonel to the Royal Marines Commandos, and it has been such a huge privilege for me to be part of the Commando family.

The commission came off the back of our first Arctic RIB expedition for which I had managed to secure the Royal Navy as one of our sponsors. We flew the White Ensign proudly from that small boat during our crossing, and through the expedition and subsequent press we shone a small light across the adventure community on the incredible spirit and values that the Royal Navy epitomizes.

The First Sea Lord at the time, Admiral West, then invited me to become an Honorary Royal Navy Lieutenant-Commander, and to represent the Navy in many other endeavours. I served this role with great pride for five years.

But at one of their 'review' boards, the Admiralty team noticed that almost every naval event I had done had also somehow managed to involve the Royal Marines. They smiled when recounting this, suspecting my guiding hand on what events I was 'available' for, and I couldn't disguise my bias to all things Commando. I was duly asked if I would like to swap my commission from the Navy to the Marines, and to become their Honorary Colonel instead. I knew it was the right thing to do, and accepted.

The fact that my father had also been a Royal Marines officer was definitely a factor behind my enduring love of the Commandos, and as they say: once a Marine, always a Marine.

My own first exposure to the Commandos had been down at Lympstone, at CTCRM (Commando Training Centre, Royal Marines), aged sixteen. I had travelled down there to attempt to complete the Potential Officers Course. I remember being the youngest person on the course, and in a very competitive field – and looking back now, I had been maybe a little lucky to pass at such a young age.

But pass I did, even if by the skin of my teeth. (Then again, is there any other way to get through Royal Marines training?) In my favour, though, I had trained hard for that course. My old school friends often remind me to this day of the 'crazy mud runs' that I used to go on after lessons each day. But my goal had been to get Marine fit, and it had paid off.

To go back to CTCRM now, all these years later, as their Colonel, means so much to me. And I love visiting and chatting with all the young recruits, especially those in their first few weeks of training. Always wide-eyed but super keen, and with hands as rough as sandpaper. It's a common trait down at CTCRM, for sure.

I recently ran part of the 30-miler final Commando test with some of the young Marines as they neared the end of their training, and it reminded me of two very stark truths. First, these guys are tough, and second, it all gets harder as you get older. It took me a week to be able to walk properly after it.

I will also never forget taking my first ever Passing-out Parade down there for the young Marines completing their nine-month training. Being kitted out in full No. 1 Dress, with the CO's sword at my side, and walking down the steps from the Officers' Mess to the Parade Square was surreal to me. I remember sensing the palpable presence of my father beside me. He would have been so proud.

The Parade went to plan – although my marching was definitely

283

off step on occasions – and as I stood before them and spoke, I knew what I wanted to say. 'You are joining the elite, a family that will stand beside you through the storms of life. The cost of joining has been a hundred barrels of sweat. But worthwhile things in life never come easy. Your journey now begins. Look after your oppo [your buddy beside you], walk humbly, and let the Commando values be your guide: courage, determination, cheerfulness in adversity and selflessness.'

Those are great values to live by.

Soon after this I was asked to visit the Royal Marines embarked on the new aircraft carrier *Queen Elizabeth*. The ship was positioned off the east coast of America and was conducting the first trials on board of the new F-35 stealth fighter jets.

I was travelling and filming with Del at the time, but as he is a former US Air Force Sergeant I figured I should try to get him security clearance to join me on the carrier. It worked, and he was soon on the roster, so we cleared a few days in the schedule to get out to visit HMS *Queen Elizabeth* in person.

As we put on our survival suits for the flight over from the US naval base near Washington DC, I looked across at Del and smiled. A friendship formed through so many moments like these. Good times.

We took off in one of the new Royal Navy Wildcat helicopters and flew low-level across the sea towards the huge carrier. Sat there, with our feet dangling over the tailgate as the chopper skimmed above the ocean, was amazing, and I kept thinking about what a journey it had been from that first time I had gone down to the Marines aged sixteen. It's sometimes hard to explain life's directions and turns.

It was a special few days, touring the ship's company and meeting the Commandos on board, and then watching the jets conduct their first landings and 'dummy' bombing runs from the tower. It was all incredible to be part of. A privilege.

The ship itself was mind-blowing in scale, and we got to stay in the Captain's guest cabins, and to meet so many of the crew, from medics and mechanics to gunners and chefs. All dedicated, humble, hard-working professionals. Examples to me, and more of the true heroes in my life.

After departing the carrier, we soon landed alongside Air Force One in the secure compound of Andrews Air Force Base. I looked across at the huge hulk of the presidential plane, so cool to see close up. I remembered our adventure with the last President, POTUS44, President Obama. How surreal that had been, too. And then I thought of the many other adventures done and dusted since. I had to pinch myself. How had all this ever come about?

The irony was that within a week of returning ashore we got a call from Discovery Channel to ask how I would feel about taking another world leader into the wild. This time, the leader of an emerging super-power with a population numbering a billion more than America's.

We were about to go on an adventure with the Prime Minister of India.

79

WINNING FRIENDS AND INFLUENCING PEOPLE

INDIA DESCRIBES ITSELF as the world's largest democracy. That's quite a power position to hold.

But what I love most about the country is that a sense of chaos appears to reign, yet somehow, through persistence and collaboration, and lots of holding hands (as is the custom), they get there in the end. In other words, they've figured out what really matters.

That kind of sums up how our Prime Minister Modi episode went.

Once again I found myself on the edge of a forest, getting eaten by mosquitoes, nervous as hell, waiting for the leader of a global superpower to appear, as if by magic.

Only this time, Mother Nature wasn't playing ball.

As so often in my life, before the beginning of some of our biggest moments we have had some intense opposition. It is as if before we can get to the good stuff, we have to prove our mettle and endure a few storms. In this case, a physical one. And it was torrential.

The PM's helicopter, laden with his security detail and advisers, couldn't get to the predetermined LZ in the heart of the Uttarakhand rainforest in northern India. The conditions were too bad.

We went to plan B.

B for boats.

The PM and his team landed at the secondary LZ and then loaded into small aluminium boats with even smaller outboard engines – a far cry from the type of thing you might imagine a leader of his standing and influence to be travelling anywhere in.

Then they started to make their way upriver.

Half of our crew were with the PM team, so they could capture his arrival, and the other half were with me at the planned rendezvous 3 miles upriver. The point where we would meet officially, and our adventure would begin.

The wind and rain started to pick up.

The Secret Service umbrellas were all up, but the PM had said he didn't need one. He was doing just fine. He was obviously determined to embrace the adventure. But our team were starting to worry that the real adventure wouldn't ever happen. Meanwhile, water was cascading over the sides and into the small boats. Everyone was bailing like mad.

The wind was slowing the boats' progress to a crawl and the waves were getting bigger as the river narrowed. If these boats sank, God only knows what the protocol would be then. Our team exchanged worried glances.

And all the while we were burning into our shoot time, and into a schedule that was already crazy tight. Even by our standards.

There was a palpable sense of tension from everyone now. We were good and we were fast, but unless their schedule was going to get shifted, it was looking almost impossible that we would get to have any real time with the leader of India.

The PM was due in a few hours at a huge rally, with some hundred thousand people waiting for him in a stadium about forty-five minutes' flight time away. But there was no way, with this added river leg that had been going on for almost an hour at this point, that we would have time to shoot together and then get the PM back to his heli in order to fly out on time.

Regardless, the PM's team kept saying sorry but they had to stick to the plan.

The moment I remember best was waiting upriver, on radio

comms to the rest of our team, and seeing the plan slipping away from us in real time. We had come so far and were now so close, but it seemed that the plan had failed, and the boats would have to turn back before we got to do anything.

Rob Buchta, my great buddy who co-produces so many of our shows out on the ground with me, a man who is as relentlessly upbeat, humble and determined as anyone I know, came on the radio and simply said, 'I know you always say never give up, Bear, but hey, I'm out of options on this one.'

'Just keep the boats upright, keep bailing water, and just get him to the start line,' I replied.

'But then what? His team are saying they can delay thirty minutes at max, and even we can't shoot an episode in thirty minutes.'

'Just get him to us and I'll figure something out . . .'

I knew that once we had met, with all the fun and energy that is always there at the start of any *Running Wild* journey, I could keep him, his team and their enthusiasm alive for at least a couple of hours. And in that time we could pull off our episode.

'And the rally?' Rob came back on the radio.

'The rally might be tricky for him to make on time,' I agreed, 'but let's not worry about that part. Just get him here, however you can.'

I looked around at Stani and Mungo and the team.

'You always say you want to bend time on these shoots,' Pete, our soundie, added drily. 'Time to work the time warp, BG.'

We had done so much prep work to pre-agree a route and a plan with the Secret Service team, but I knew now that none of that was going to work. We were running out of time too fast.

'Ideas, team?'

Mungo looked at me and said, 'I reckon just ditch the planned climb up that lookout tower. In fact, ditch everything. Just meet the dude, walk off into the forest, and we'll follow. Improvise – you're good at that.' He then smiled at me.

That same Mungo who had been with me in the heat of the Sahara with the French Foreign Legion all those years earlier. My first ever TV series. He was still here. Still doing the hard part. And still smiling through the storms. A friendship that has endured. Back then, in the desert, as a man of faith, he had quietly offered to say a prayer for me before we'd started filming. He knew that I was nervous. I never forgot that moment.

Mungo then added, 'Anyway, just seeing you and the PM like drowned rats in this pouring rain will look amazing.'

80

THE WILD DOESN'T CARE
WHO YOU ARE

BEFORE THE SHOOT, we had been going back and forth with the PM's planning team, trying to find the right time in his schedule to film together. When the call had finally come through saying that we were 'Green on . . . Go', I had been doing a press junket in Los Angeles for our natural history show *Hostile Planet* and for *Running Wild*, both on National Geographic.

I was super happy to have a good excuse to leave early. I packed up fast, got on a plane to Europe, then on to Delhi. Then from the airport a chopper flight into the mountains, followed by a long 4×4 trip deep into the forest.

I finally arrived at 3 a.m. at a small lodge that the crew had been using as a production base. I was tired, but adrenalized.

The team had planned a simple journey for the PM, and it had all been signed off days earlier. But that plan had now gone the same way as so many best-laid plans. It had turned to mud.

To be able to thrive in uncertainty takes practice and experience, and also a winning team supporting one another through a crisis. We had all those elements. It was time to use them now.

The radio crackled. It was Rob.

'On our way to you. Two mins out. Then it's over to you, BG.'

I saw the PM's 4×4 motorcade appear through the forest, having driven the final mile or so from their river arrival point.

Time to go to work.

From the moment he stepped out of the jeep smiling, I knew we would be OK. We shook hands, I gave him the briefest of briefings on the new plan, and before anyone could say a word against it I led the PM off into the forest, with his close protection team bustling along behind with their weapons at the ready.

I soon got the PM to help me make a simple improvised spear by strapping my knife to a stick, as a precaution against tiger attack. We were definitely in tiger country, and as ambush predators the threat was justifiable, but the PM was at pains to say he would never kill a tiger.

I agreed, obviously. Still, protection is protection.

'Maybe give me the spear then, sir?'

'No, no, I got it,' he replied, grinning.

I think he just liked holding the thing. I mean, who wouldn't?

And off we went.

My worst moment was probably asking the Prime Minister of India what it felt like to be the most powerful man in China.

'India, you muppet,' Mungo chuckled from behind his lens.

I winced. Such an embarrassing mistake. Modi just smiled warmly.

As the rain got worse, it became harder and harder to hear the translator on my hidden earpiece. The plan had been for the PM's simultaneous translation lady to be close by, speaking into her microphone directly Bluetoothed to me. But the static of the storm meant the feedback in my ear was too difficult to hear through.

I took the earpiece out and slipped it into my pocket, before looking across to Pete, our soundie, with a nod.

A few minutes later, the lady tapped Pete on the shoulder.

'I really don't think Bear can hear me.'

'That's maybe because the earpiece is in his pocket.' Pete smiled. 'But don't worry, he'll muddle through.' Pete, always the optimist.

It was definitely a little hit and miss when it came to understanding much, but I kept asking the PM the questions I'd planned to ask.

Apparently, there was a classic moment back in the edit in Los Angeles afterwards when the Hindi translation of his words was being laid in and matched to mine. The PM had been busy telling me about the need for infrastructure in India in order to empower economic growth across the country. I had paused, looked at him intently, then with a totally straight face had simply said, 'Yes sir, I agree entirely. Tigers can indeed be very dangerous at this time of year.'

You can't win them all.

To his huge credit, the PM was game throughout, and he laughed and spoke with insight and humility about his own life, and the journey he'd taken to get to where he was today. We discussed the importance of India becoming a global leader in terms of conservation and protecting our planet. The PM spoke about how he wanted to change the way India operates, in terms of fossil fuel dependence and the protection of wildlife and wild places.

We finally made it down to a tributary of the main river, to where the crew had already partially constructed a small and highly improvised 'doughnut' raft from some reeds and an old tarpaulin.

One of the only prior stipulations we had received from the Secret Service team when planning the journey was that under no circumstances could the PM get his feet wet.

I looked at the PM now, drenched to the bone, and considered that as he was now soaking wet anyway a quick river crossing in this raft wouldn't harm anyone.

I knew it would be iconic, to have one of the most powerful men on Earth in a little coracle made of reeds on a jungle river in a storm. I knew that it would speak to the fact that the wild doesn't care who we are, that it treats us all the same – and that it bonds us all in tough moments.

I was determined to pull this river crossing off and gave the PM another of those brief briefings. And then we launched this little coracle into the river.

As the PM put his weight in, the raft started to sink. I knew there was no way it would hold us both.

The Secret Service were suddenly getting very nervous at all this as they gathered behind the camera alongside us on the bank.

Keep pushing, Bear, I thought to myself, *just get this thing out into the main flow and then we are free . . .*

I still consider it one of the most special moments in my life. To be out in that river with the PM, just us two, in the torrential rain, barely being able to understand each other's words but just laughing at the moment. It is the heart of what I love about the wild. It is always the great leveller, and connector.

We finally reached the far bank and sat on the stones, both drenched and cold now. I covered him in a small poncho to try to keep him warm. Again, connections.

The PM had also seen the Obama episode, and had specifically asked if he could pray at the end of this journey, like I had with the US President. I definitely hadn't seen that one coming. But he prayed, and it was beautiful. Then, just like that, we were out of time. But we knew we had our episode. And more.

The PM team bustled him into one of the jeeps and they were gone. We almost couldn't believe what we had got. Rob turned to me and we hugged.

'Good job holding the ship together today,' he said. 'And I don't just mean the raft!' We both laughed. 'I definitely felt it was all starting to slip through our grasp at one point, eh?'

'We did it, brother. Epic team.' I paused. 'You reckon they'll do us some pina coladas back at base?'

We never found out the answer to that question as we ended up having a marathon sixteen-hour road trip back to Delhi. We were all whacked, but also so proud to have pulled off what we had. It's a true testament to the ability of our team to adapt, to be able to pull off this stuff and always to make it fun and empowering, often against the odds.

Tragically, the day was later overshadowed by us hearing the awful news that a terrorist attack had taken place in Kashmir. The timing was such that, literally as we wrapped our journey together, the attack had been reported.

What I do know is that if that terrorist attack had occurred just an hour earlier, our mission together would certainly never have happened.

That journey with the Prime Minister led on to us filming many more episodes with some of the biggest stars in India, such as Akshay Kumar and Rajinikanth – and that's been a great privilege indeed. I see it as the start of a wonderful friendship with a country I had never expected to fall in love with so much.

Beautiful hearts and wilderness.

81

SCAR TISSUE

SOMETIMES I GET annoyed at how much it hurts to get up in the morning nowadays. My body aches and old injuries never seem to get any better. But a little shake down, a few stretches, a healthy breakfast and I am generally good to go again.

I love exercising as part of my day. It's now a lifelong habit that I won't ever let go of. Part of the day. It's never for long, just thirty minutes max, mixing up strength and cardio and flexibility. After my workout, if I'm at home, our boys generally know where to find me. We have an old iron cold-water farm trough that is brutal in winter. But whatever the weather, wind, rain or snow, I love to get in it.

Jesse has also taken to joining me in the trough these days, and it makes for some of my favourite father-son moments. A shared dip of horror.

It's that ethos of doing a little bit of 'internal wintering' and resilience training every day – seeking out the uncomfortable. It doesn't have to be for long, but I know that a small daily dose of hardship, whether that's a killer workout or an ice-water immersion, is good for me. It's about keeping that 'never give up' muscle tested and strong. The most important muscle we all possess. Then, when I'm back out doing the day job, and charging around the mountains, I am ready. Inside and out. It's about doing all we can to stay prepared.

There is no doubt that both age and injury keep all of us humble.

You can't run from those things. I try to accept them as best I can, but I'm not doing a brilliant job, to be honest. I still consider both as battles to be fought. But I will keep trying. And adapting.

The vertebrae around where I broke my back are as stiff as a board. The scar tissue around T8, T10 and T12 is like a ramrod. Solid. That injury never lets me forget it's there. Or the busted toes, the dislocated shoulder, the mashed-up quadricep, or the sun-damaged retinas . . . all gentle reminders of some epic adventures.

The one part of me that has held up remarkably well are my knees. Considering the number of ex-SAS soldiers I know whose knees have been damaged from carrying heavy weights for long periods in the mountains, or from parachuting at night and landing blind, I am doing OK.

Then there are the remnants from frostnip in both my feet and fingers. Everest left me with that – lest I forget her – and the pain nowadays of rewarming them, after a thousand ice lakes on *Man vs. Wild*, *Running Wild* and *You vs. Wild* shoots ever since, is simply the price of that Everest ascent.

So all in all, I guess I mustn't grumble.

And as for the odd minor injury, like my busted, bent nose, I see it simply as added character.

Nerve damage is the one thing that is both annoying and painful – especially in my finger, from the time I almost sliced off my entire digit in a Vietnam jungle. Grabbing a razor-sharp shard of bamboo and trying to pull it out of the ground was dumb. The bamboo was always going to win.

The Vietnamese hospital we finally located in some remote town was another solid reminder never to complain. There wasn't a lot of anaesthetic action going on around me in the dirty A&E section there, yet no one seemed to be making much of a fuss about it.

If we're not on the breadline, and we have a roof over our heads, and access to health care, then that makes us all pretty lucky, really.

There are tough people everywhere, living tough lives. The media just doesn't always celebrate them.

When I look at the bigger picture like this, and remember the list of my own near-misses, it is a humbling reminder of how fortunate I am – just to be alive.

After all, historically, if you lived much past thirty-five you could consider yourself a real survivor.

82

THE KING'S CUP

MY MOTHER STILL lives in the same house that I grew up in, in Bembridge on the Isle of Wight, a small island off the south coast of England. After Dad died so young, it is amazing how Mum has kept living so fully and so positively. She is an inspiration and a role model for embracing life, and it is why she is loved by so many in her community.

Note I say how my mother lives life fully, not necessarily conventionally, because Mum is unashamedly original. Try as Lara (my sister) and I have done over the years, Mum won't be told how to live – even if it isn't always doing her much good. Then again, who are we to say what is right and wrong? Mum loves to live in one room downstairs in our old house, with rarely any heating. She moved her double bed into the sitting room and there she resides, surrounded by piles and piles of letters, notes, newspapers, photos – and cats.

Her garden is full of pigs and all sorts of animals that surprise her many visitors, and the whole place often becomes so dirty that you might look at it and wonder if she's all right in such a mess.

We try and try to get it cleaned and tidied up, but she hates it and it never lasts for longer than a day in any sort of respectable state. The cats lick the food on the kitchen table and it is not uncommon to find cans more than two decades past their sell-by date. Mum, by her own admission, calls herself a hoarder, but she is also undeniably happy and positive, all the time. I guess that being yourself means no peer pressure. And in society nowadays that's a rare and beautiful thing.

Mum is now seventy-nine and moves around the village on a half-broken mobility scooter and uses a twenty-year-old car for longer journeys. She often texts me after she has taken her mobility scooter out in the middle of the night down to the sea to watch the moon on the water. She says she likes the night-time as it is so quiet and she can feel God's presence. Who am I to argue with that?

Mum refuses to let me replace the scooter or her car for her. She says she likes being able to see the road through the floor of the footwell and that old things are like old friends. They are part of who she is.

Mum is also impossible to reason with.

Last summer, I was invited by Prince William and the Duchess of Cambridge to skipper a boat for them in a race during the Isle of Wight's annual Cowes Week yachting regatta. Only 9 miles from where Mum lives and from where I grew up.

The royal family were planning a fundraising day for their chosen charities, and Prince William wanted to re-establish the awarding of an incredible gold trophy that his great-great-grandfather, George V, had first presented in 1920.

The only issue was that the race was bang in the middle of our precious island time together as a family, and as you know by now we always do all we can to protect that time away. That means saying no to all invitations. But this was special and, just as importantly, I knew Mum would love me to do it.

Now, Mum is a true monarchist, and there are as many photos of royal babies and weddings around her bed as there are of the family. She was up in arms just at the thought that we might not accept such an invitation.

'After all,' she said down the telephone, 'if your monarch summons you, like this, it really would be so churlish not to accept.'

I am yet to figure out what 'churlish' actually means, but Mum always levels it at me whenever she wants me to do something. And it generally seems to work.

That aside, I felt it would be a great chance to see Mum, even if it meant a long journey down south from North Wales.

The race would have eight yachts, with eight 'personalities' to skipper them, and with the King's Cup at stake for the winner. Prince William and the Duchess of Cambridge would also be skippering a boat each. I made a quiet note to myself to make sure I didn't beat either of them. I didn't think that would be good protocol.

As it happened, on the day we were all due to fly as a family down south, the weather was horrendous. It didn't bode well for either the race or for getting a helicopter on and then off our island. Shara isn't a great flyer at the best of times, and choppers in 60mph winds landing on cliff tops in driving rain are not her favourite. So in the end we decided that I would go solo, and make it a fast in-and-out trip, so as not to leave the family for too long. That way I would still get to do the race and see Mum, and be home in under twenty-four hours.

The only clear weather window meant a dawn mission south and we just managed to get to the Isle of Wight before the storms hit. The roads around Cowes are always busy anyway during that regatta, but throw in the royal visit and the heightened police presence and the place was rammed. The Land Rover sponsorship team picked me up and drove me the last mile down to the quayside.

As we came down the hill and wiggled our way around the cordoned-off streets, I caught a glimpse of a lone mobility scooter pottering along at 5mph down the no-access road that led to the Royal Yacht Squadron, where Prince William and the family were due to meet us all any minute. When I saw a flash of kaftan flowing behind, about to be caught in the wheels at any moment, I knew at once who was on that scooter.

The police all looked at each other a bit confused. They obviously knew this wasn't a security threat but seemed unsure whether it was best to wrestle this old lady to the ground or let her through. I leapt out of the car and tried to chase after her. I was dying inside. This was going to be so embarrassing.

I failed to reach her before the scooter disappeared around a corner. All I could now see were hundreds of spectators looking at each other, as if wondering who on earth this eccentric lady on a mobility scooter was, with a brightly coloured kaftan flowing behind her like some elderly Victorian empress.

At least I knew Mum had now made it to Cowes safely.

I then got a text from the race organizers to say that Mum had made it in time, just before the royal party, and that she had met up with Lara who was also down on the Isle of Wight at that time. They said she was all ready to watch the race from the Castle of the Royal Yacht Squadron.

Good. That's one thing sorted, I thought to myself.

The wind was due to drop slightly for the next few hours, so the race was suddenly brought forward at a moment's notice. It was a rapid 'all move' to get to the boats and to start the race.

There was little time for any training or formalities, and I quickly changed into waterproofs and leapt aboard the Fast-40 yacht. She was an incredibly sleek, dynamic boat, and looked amazing all branded up to show our support of the TUSK charity that supports conservation in Africa. This would be fun.

I had managed to work it so that I was skippering the boat with the one crew I already knew. We certainly weren't the favourites but I knew we would have fun, and most importantly it was the team that included my buddy John Coffey, who had been on our North West Passage Arctic RIB expedition several years before. Ten seconds of hellos and then we cast off, spray ripping across the deck. I could already sense that this was going to be a fast and furious race.

It was amazing to be skippering a racing yacht of this calibre, and even though I had been brought up on the waters down here, sailing most days, I hadn't ever skippered anything this fast. It was exciting, loud and wet, with both high adrenalin and high stakes.

The start line was a flurry of eight yachts weaving in and out of each other at high speed, plus the official press and race boats, all trying to avoid each other in ever-worsening conditions. I was loving it.

I was fortunate in that most of the other skippers weren't sailors, so they were firmly in the grip of one of the lead crew to do exactly as instructed. But in fast conditions, that relay of information means seconds lost, which in turn gave me a natural advantage.

Before I knew it we were in the lead, and all of a sudden I was determined we were going to win this race, if not just for TUSK, then for my mum.

I rapidly forgot I was racing the future King and Queen.

83

STORMS MAKE US STRONGER

THE EXHILARATION OF victory and the celebrations of our crew were special to see – it meant a lot to them, and we had won against all the odds. The TUSK charity team, on the shore, came on the radio, totally over the moon. The press boats surrounded us and the crew lapped up their moment.

For me, winning isn't necessarily the fun part. The moments I really thrive on often happen unseen, and are gone before we realize it. In that race, they had come early on, as we scrapped and scrambled to gain our lead, always fighting against the big waves and strong winds. Those are the moments I love, whether in a jungle or on a mountain or out at sea.

But on this day, I was just so proud to win that race for my mother.

When I was growing up on the Isle of Wight, both my parents had always helped and encouraged an adventure spirit in me. Whether that meant allowing me to tinker around building a little wooden boat in the garden or scrambling along the sea cliffs on the headland near our home. It was always about going for things in life, and understanding and respecting the power of nature.

I remember those school holidays so vividly as a young boy helming my little dinghy far from the safety of the small harbour near home, and battling ever-mounting waves and strong winds, staying out at sea long after others had returned to harbour. I loved those battles against the elements. They made me feel that potent mix of being truly alive and alert, combined with an intense trepidation, and having continually to adapt to the unexpected, whether it be a squall or a crumbling chalk cliff.

Those times taught me simple lessons, learnt accidentally and almost always through failure and experience. I look back now and know that subconsciously they have helped me so much in my career. Plus, of course, they gave me an ability to stay calm in storms.

I didn't realize any of this back then, but I see it now. That upbringing down on the Isle of Wight, where in the long winter months there were few people around and I had to make my own fun, was the time when I found my adventure spirit.

It was also the only real thing I was ever much good at. At school I was pretty average at sport, and pretty average in the classroom, but back home I was becoming a ninja at climbing trees, scaling cliffs, making stuff and then testing myself on the sea. When it came to surviving in Mother Nature, I was almost accidentally becoming quite good. I had a quiet confidence at a young age in being competent outdoors that I certainly didn't have in other areas of my life.

During the many sailing races and teaching weeks that I did growing up, the bits I loved most were those moments when the rescue boats were sent out to bring us all in, when the weather had turned. Those were the times I really came alive. The bigger the waves, the stronger the rain and the more violent the squalls, the more it excited me. Bring it on. I can feel that excitement inside me right now, as I write.

The races were OK, but I never won much. I didn't have a massive competitive drive to beat others. My drive was to test myself and push my boundaries. Winning over Mother Nature was much more exciting to me than any medal. (Although, time soon taught me that we don't ever beat Mother Nature, we simply learn the skills and attitudes needed to survive her.)

The heart of all this is that at a young age I learnt that the storms we face, and every time we survive them, make us more resilient. It's that adage that whatever doesn't kill us makes us stronger.

Each time I came off the sea having made it back to shore after a squall, every time I scrambled over the top of the sea cliffs with Dad, against the odds, and despite my fears looking down, every time in winter that I went hill running in heavy army boots that were too big for me, having hosed myself down first so that I was soaking wet, was an inner test.

I might be exhausted physically, freezing cold, weather-beaten, with wrinkled-up fingers, but every time I felt stronger. Inside and out.

84

EATING ANTS WITH THE FUTURE KING

BACK ON DRY land, Prince William and the Duchess of Cambridge seemed to take great pleasure in ribbing me about beating them. They always conduct themselves with such grace, and the more I see them in action or spend time with them at various projects, such as with the TUSK conservation awards or at Scouting events that they have helped support, the more their true class shines through. And true class has nothing to do with inherited titles but all to do with humility, kindness and dedication to support those who need it most. When it comes to those qualities, they are in a league of their own.

As I was heading to the official prize-giving after the race, a member of staff came up and asked if I would say hello to Prince George and his friends, who had been watching and wanted to say hi.

Prince William had previously told me with a smile that their children all loved watching our Netflix interactive show *You vs. Wild*, so, of course, I agreed to go round the back of the Royal Yacht Squadron Castle, to meet the future King.

As I stooped down to shake Prince George's little hand, I noticed a trail of ants marching across the paving in between us. Prince George then stared down at them too, then looked up at me, eyes wide with trepidation and excitement. I knew that look well.

I smiled at him and said, 'Have you ever eaten an ant?'

He shook his head nervously.

'You want to? Just you and me?'

He looked around, then back at me, and nodded.

I grabbed one ant for him and got him to hold it between his little fingers, then got one for myself.

'One, two, three, and in it goes . . . ready?'

He looked more nervous now. I smiled reassuringly.

'It'll be fine. We can do this . . . OK? One, two, three, here we go!'

I will never tire of that wonderful grimace on people's faces as they eat something from the survivor menu for the first time. Whoever they are. It's always priceless. But I don't think I will ever see wider eyes than Prince George's, nor a broader grin on completion of 'down the hatch it goes'.

It was a fun moment that I hope when he is King one day and I am an old man he might still remember. After all, who can ever forget eating their first ant?

The real point of this whole King's Cup story was to get you to this next one moment, which for me sums up so much of my life so far.

The race day ended with Prince William, the Duchess of Cambridge and me on the outdoor stage of the Race Regatta, thanking the charities and the many folk who had helped make the event happen. Then it came to the awarding of the actual cup itself, and it really is something to see. It must be half a metre high and made of solid gold.

I stepped across to receive it from Prince William and then turned to thank my crew of the day who had been the real heroes of it all. Then I looked down from the stage towards my mother, sat in her mobility scooter, next to my sister Lara. They were right there at the front of the crowd, where I had managed to get them seats. Mum had been so excited just to get to see Prince William and the Duchess of Cambridge close up.

As I tried to keep a hold on this massive cup and wrap up my thank-yous, I could see tears falling down my mother's cheeks. I found myself suddenly lost for what I was meant to be saying.

All I could see was two people who I love more than words will ever say; the two people, along with my late dad, who when I was growing up had helped me so much and believed in me when I wasn't particularly outstanding at school in any way. To my mum, I will always be the son she had after having endured so many back-to-back miscarriages, and to my sister Lara, eight years older than me, I will always be her baby brother whom she treated like a teddy to throw around and look after.

I realized in that moment that things like cups and awards and applause, and all the stuff that we often get wrapped up in chasing, for whatever motivation, are actually the least important part of life. The best stuff, the real stuff, is always going to be found with those we love and those closest to us.

I had to choke back a surge of emotion. But then all of a sudden I knew exactly what I wanted to say.

I dedicated the cup to my mum and to my sister, saying how I owed them so much. I remarked how strange life can be sometimes. That here I now was, on the island where I took my first little adventure steps as a kid, and to have my mum and sister still next to me, still supporting and encouraging, meant so much.

I reminded them how I had never won a race on the island before, but that I'd learnt to challenge myself – to thrive and love the outdoors in this very place. And that the lesson to me is always keep giving, keep trying – chase the right stuff not the wrong stuff.

Don't worry about the results or the races, but embrace the challenge and the testing of spirit, that internal effort muscle. Because the irony is that when we chase the right stuff, with determination, for long enough, the cups and awards will often follow.

I managed to give Prince George a final wink, and thanked the event sponsor, Land Rover, for their support of me and our team, backing us on so many adventures over the years. Then I went down to give Mum and Lara a hug. Both are such larger-than-life characters, much more extrovert than me, but family bonds are hard to break, and it meant so much to share it with them and to see Mum so happy.

Early the next morning I was back on our little Welsh island hide-away with Shara and the boys. Job done. I had promised the family I would be back within twenty-four hours. I wasn't far out.

It was still blowing a gale up there, solid 50mph winds, but the pilot just managed to land safely. The family were still asleep so I thought I'd get my workout in before they woke up.

Before I knew it I was in my shorts, shirt off, bare feet, doing my session in the pouring rain behind the lighthouse, swinging that rusty old 24kg kettlebell. Happy days. I did my final run down to the jetty, fifteen good pull-ups on the old scaffolding bar wedged into the crevice across the top of our seal cave, then once more back up the steep path to the lighthouse.

At this stage my heart rate is always through the roof and I'm doubled over to recover my breath. The island workouts are always brutal, mainly due to the gradient of the hill and the incessant south-westerly winds that batter you back as you lean into the 400-metre ascent.

Check watch. Twenty-nine mins. Just within my self-imposed time limit. Just. The wind always makes it a close-run thing to complete my island circuits in time.

I then sprinted back down to the jetty which was being battered by the waves, and dived into the sea. I know the island currents well, but still I am always super careful with rips and eddies, in both sea and rivers. Being alone, I stayed close to the quay and looked up at the angry skies.

There was something magical about floating in the frothing sea, in the torrential rain, no one else for miles around, family asleep back up the hill, and thinking about the last few days. I've always liked that split world. Between working and filming and the craziness that often surrounds that part of my life, and the calm, quiet, natural existence that plays such a pivotal role in our family life.

Being back on the island reminded me why I love our way of life so much. In truth, so much more than the glitz. Lara always jokes that she would be much better at being famous than me. She is probably right. She loves a bit of glamour and lots of parties. It's just her exuberant nature. I just love great adventures with good buddies and then being back at home.

Still, the last day has been fun, I thought to myself. *Especially the ants bit . . . the boys will like hearing about that.*

I smiled.

And winning the race.

I clambered back on to the jetty and headed up to the family.

85

ONLY JUST GETTING GOING . . .

I LOOK BACK now and I am so grateful for the many great adventures Shara and I have had together already – and having our three boys has undoubtedly made the whole journey that much more fun. It's hard to put into words, the joy of children, and I won't patronize you here. You get it, I am sure. But it is about something far greater than careers, or fame, or wealth, or any of that transient stuff that means so little in the real game of life.

There is no doubt that no one takes the mick out of me more regularly and mercilessly than Shara, Jesse, Marmaduke and Huckleberry. It's painful at times, but generally well founded, and in so many ways I have become an all too predictable parody of myself. But I don't mind that, and I know that life is always better when we can laugh at ourselves. And I make no apology to the family for all the many things I harp on about too much. For over-loving (if there is such a thing) all that the Scouts stand for, for over-sharing the ethos that the Royal Marines live by, for telling too often the same stories of Trucker and me while in 21 SAS – not to mention saying 'never give up' way too much and for insisting on praying for all of them in bed every night.

These are all good things.

I try to remind them also of the five Fs for a good life. Family.

Friends. Fun. Faith. Follow your dreams. Keep these front and centre and you will be doing OK.

Sometimes I pinch myself that I have somehow managed to come through so many close scrapes, not just in terms of physical danger, but also those very real early TV battles that, had I lost, would have changed the direction of our family life so dramatically.

It makes me so aware of the role luck invariably plays in all our lives. I can't quantify it, and I don't want to belittle the role of bloody-minded persistence, because ultimately that is always going to be king, but it is just that as I get older I also become ever more aware of the huge part good fortune and good timing have played.

But then again, as the champion golfer Gary Player once said, 'The harder I work, the luckier I get.' I've always liked that.

I think back to when I first applied to the Royal Marines to join as a young officer, and that Potential Officers Course when I was sixteen. I was as wet behind the ears as it is possible to be. I was truly terrified, sat on an old train as it trundled along the final few miles to Commando Training Centre, Royal Marines, Lympstone. I was dressed in a neat tweed blazer with my tie scrunched up awkwardly around my neck. I had my HM Forces rail warrant in my hand and a lump in my throat that was making me feel sick.

I glanced down the carriage at all the other recruits, who stood out a mile. They all looked so confident and strong. None of us spoke and I tried instead to look out of the window as we edged along the estuary.

At the end of that course, out of the thirty recruits, only fifteen of us were passed to continue on to the Admiralty Interview Board. And from that, only four of us were passed for training. I remember that I got through by the narrowest of margins. My final report stated 'his happy-go-lucky attitude will need to be harnessed in his career moving forward'.

Looking back, I see how that 'happy-go-lucky' streak has always been such a key quality in my life. When we live with optimism, the world often rewards that (even if, at times, the optimism is a little misplaced, or in my case a little naive). It's a fire inside that we must do everything we can to feed and encourage, in ourselves and others. Optimism has carried me through many a dark night, and I believe it to be a muscle that grows when used.

There will always be storms, no doubt, and every relationship and project at times requires grit and commitment if they are to be successful – but therein lies the magic. It means we all have the capacity to succeed, because so much of success comes down to attitude.

The other key factor for me, of course, has been an amazing team. The ability to recognize, foster and protect those friendships and relationships has been fundamental to any success. You now know many of those names that are so core to my life. And I hope they will be beside me for ever.

None more so than Shara.

When it comes to Shara, there is so much I could say. The overriding thing with her is her kind heart. Kind to me, kind to those who need it most, kind in her everyday dealings. Is there anything more important in life?

Whenever I come across an overly pushy wife or husband, someone who maybe loves the limelight quite a lot – as can so often happen – it makes me realize more and more the value of a family who are unchangeable, whether by fame, wealth or power. All those transient things can be such a drug to people, but if they are all you pursue then they will eat you up and so often turn relationships sour.

When it comes to fame, wealth and power, the danger is that they will leave you at the end with less than you started with, in terms of what really counts.

Shara has always had her feet so solidly grounded, and that has been the key to so much. She is loyal to a tee, and above all she has loved me, with or without a career. That has been the great gift in my life.

I've always thought that it is easy to be a hero at work (especially when we control the narrative, the interviews or the programme edits), but the only place it really matters is at home. That's the real place to try to be a hero. If I was to express how to live this out, it would be to try to be kind and humble, to endeavour to be brave in the big moments, and to be determined in everything. That's how I want to live in a nutshell. Even though I genuinely fail often.

If you want to find out what someone is really like then ask their wife or children. That's the true test. And the toughest one. But it is the only test I ultimately care about. The rest is work – a journey and a privilege for sure, but still just work. The real me has always been found at home. And I am proud of that.

As for our boys, Jesse, Marmaduke and Huckleberry, words don't do you guys justice. I love you more than words will ever convey, and as for all of this other weird stuff you have had to grow up with, I hope, like me, you know it's nothing – especially compared to all the fun and adventures Mama and I have with you. Those adventures are always my favourite.

And as for the future, well, we have had *Mud, Sweat and Tears*, and now here is *Never Give Up*. Who knows what will follow? That's always the great mystery for us all: tomorrow.

What I do know is that I'll commit to tackle it with that same courage, kindness and NGU spirit. And a faith that good things lie ahead. That's important too, for the harder days.

Simple things that matter a lot and that can change so much.

There is also a part of me that feels we are only just getting going. We have spent so much time and energy getting to this stage that for me it is now about what we are going to do with the platform we have that is good and lasting.

The answer always lies in other people. Encourage, encourage, encourage.

Because now more than ever, I want to be part of helping other people to find their own adventure – whether that is a *Running Wild* guest or a young girl Scout in Sudan. It is about empowering people with the tools and skills, as well as the inspiration and values, that they most need to thrive.

All of this is a journey, not a destination. Every day is a school day, as they say. And I am still learning as much as you are.

When people come up to me in the street now, they tend to say one of three things.

Either: 'Good on you with the Scouts. My son/sister/cousin was a Scout, and they loved it. Our kids are the future.'

They are right. Always build young people up. Show them by example that life is a gift that's best lived boldly, and teach them how to overcome the inevitable fears.

In America, people will often come up to me and say: 'Hey Bear, say "protein". Or "vitamins". Or "glacier". I love how you mispronounce them.' Cue lots of laughter. It's a British accent thing, I guess. I never mind it.

But the one thing I hear the most is this: 'Hey Bear: "Improvise, adapt, overcome".' Followed by a fist pump.

It's a slogan I first heard as a soldier, and one that I've used extensively on shows and expeditions over the years. I have always liked it. It feels like a fitting way to end this book.

So in that spirit, if you were to ask me about tomorrow . . . you'll already know the answer:

Improvise, adapt, overcome . . . and above all, never, ever, give up.

'The Lord stood at my
side and gave me strength.'

2 Timothy 4:17

PICTURE ACKNOWLEDGEMENTS

MANY OF THESE photos have been taken by members of the BGV team, friends and collaborators over the years. A special thank you to Ben Simms, Ben Kenobi, Boon and Emma Myrtle for so many of these images, and to the following studios and photographers:

Running Wild images (pages 1 and 3–5) courtesy of NGC Networks, US LLC, Bear Grylls Ventures, LLP and Electus LLC. *Animals on the Loose* photograph (page 2) courtesy of Netflix Studios, LLC and Netflix Global, LLC. On the radio to the team (page 6) courtesy of Potato TV and ITV Studios.

With Prime Minister Modi (page 9) courtesy of Discovery Networks Asia-Pacific PTE LTD. *Running Wild in China* (page 10) courtesy of Yunji Media (Shanghai) Co., Ltd., Bear Grylls Ventures, LLP and Electus LLC. With the Royal Marine recruits (page 11) © Crown copyright. Spending time with the Scouts (page 12) © The Scout Association/Dave Bird. With the Duchess of Cambridge at the Queen's Scout ceremony (page 12) © Olivia Harris, WPA Pool/Getty Images. Winning the Kings Cup (page 12) © Andrew Matthews, WPA Pool/Getty Images. St Tudwal's Island West (page 14) © Sara Donaldson. *GQ* cover (page 16) © The Condé Nast Publications Ltd. Always look up (page 16) © Hamish Brown.

ABOUT THE AUTHOR

BEAR GRYLLS OBE, has become known worldwide as one of the most recognized faces of survival and outdoor adventure.

Trained from a young age in martial arts, Grylls went on to spend three years as a soldier in the British Special Forces, as part of 21 SAS Regiment. It was here that he perfected many of the survival skills that his fans all over the world enjoy, as he pits himself against the worst of Mother Nature.

Despite a free-fall parachuting accident in Africa, where he broke his back in three places and endured many months in and out of military rehabilitation, Grylls recovered and went on to become one of the youngest climbers ever to reach the summit of Mount Everest.

He starred in seven seasons of the Discovery Channel's Emmy Award-nominated *Man vs. Wild* TV series, which became one of the most-watched shows on the planet, reaching an estimated 1.2 billion viewers. Since then, he has gone on to host more extreme adventure TV shows across more global networks than anyone else in the world, including six seasons of the global hit TV show *Running Wild with Bear Grylls*.

Running Wild has featured Bear taking some of the world's best-known stars on incredible adventures. These include President Obama, Julia Roberts, Roger Federer, Will Ferrell, Channing Tatum, and Kate Winslet, to name but a few. Bear also filmed an adventure with Prime Minister Modi of India, which achieved a landmark record as the 'world's most trending televised event, with a staggering 3.6 BILLION impressions'.

Bear has won two BAFTAs, including one for his hit Channel 4 show *The Island with Bear Grylls*. He also hosts the Emmy Award-nominated interactive Netflix show *You vs. Wild*, which was one of Netflix's most-watched shows in 2020. Netflix and Bear then forged new ground by making two interactive movie versions of *You vs. Wild* including the second-most-watched film on Netflix US, titled: *Animals On The Loose*.

His autobiography, *Mud, Sweat and Tears*, spent fifteen weeks at Number 1 in the *Sunday Times* bestseller list and he has now written over ninety books, selling in excess of nineteen million copies worldwide.

An Honorary Colonel to the Royal Marines Commandos, Bear is also the youngest ever UK Chief Scout and the first ever Chief Ambassador to the World Scout Organization, representing a global family of some fifty million Scouts.

He is married to Shara and together they have three sons: Jesse, Marmaduke and Huckleberry. They live on a houseboat on the Thames in London and a private island off the Welsh coast.

Bear's life motto is simple: *courage and kindness . . . and never give up!*

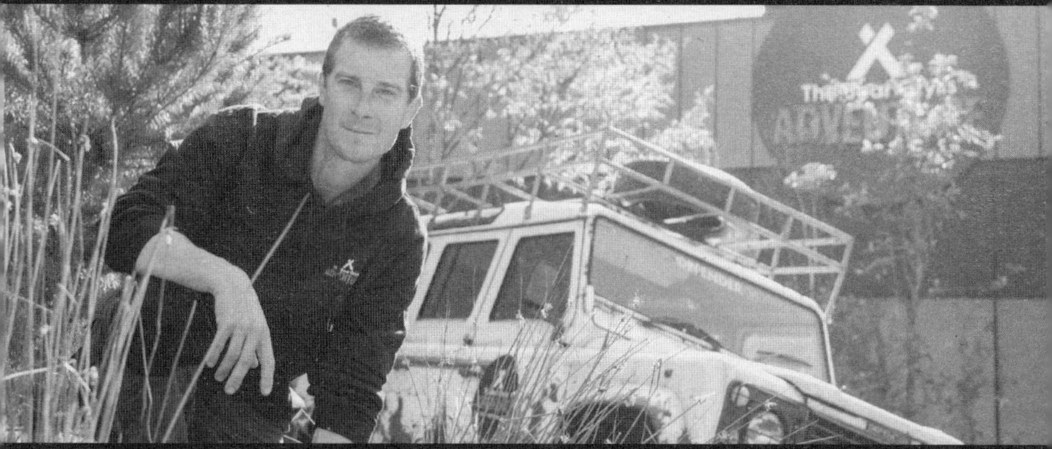

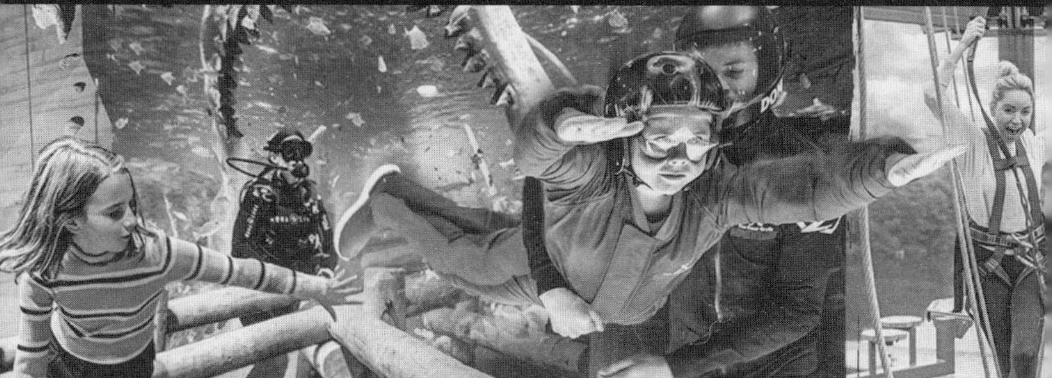